T0351305

A Student's Guide to Maxwell's Equations

Maxwell's Equations are four of the most influential equations in science: Gauss's law for electric fields, Gauss's law for magnetic fields, Faraday's law, and the Ampere–Maxwell law. In this guide for students, each equation is the subject of an entire chapter, with detailed, plain-language explanations of the physical meaning of each symbol in the equation, for both the integral and differential forms. The final chapter shows how Maxwell's Equations may be combined to produce the wave equation, the basis for the electromagnetic theory of light.

This book is a wonderful resource for undergraduate and graduate courses in electromagnetism and electromagnetics. A website hosted by the author, and available through www.cambridge.org/9780521877619, contains interactive solutions to every problem in the text. Entire solutions can be viewed immediately, or a series of hints can be given to guide the student to the final answer. The website also contains audio podcasts which walk students through each chapter, pointing out important details and explaining key concepts.

DANIEL FLEISCH is Associate Professor in the Department of Physics at Wittenberg University, Ohio. His research interests include radar cross-section measurement, radar system analysis, and ground-penetrating radar. He is a member of the American Physical Society (APS), the American Association of Physics Teachers (AAPT), and the Institute of Electrical and Electronics Engineers (IEEE).

A Student's Guide to Maxwell's Equations

DANIEL FLEISCH
Wittenberg University

CAMBRIDGE UNIVERSITY PRESS
Cambridge, New York, Melbourne, Madrid, Cape Town, Singapore,
São Paulo, Delhi, Dubai, Tokyo, Mexico City

Cambridge University Press
The Edinburgh Building, Cambridge CB2 8RU, UK

Published in the United States of America by Cambridge University Press, New York

www.cambridge.org
Information on this title: www.cambridge.org/9780521877619

First published 2008
8th printing 2010

A catalogue record for this publication is available from the British Library

Library of Congress Cataloging in Publication Data

Fleisch, Daniel A.
A Student's guide to Maxwell's equations / Daniel Fleisch.
p. cm.
Includes bibliographical references and index.
ISBN 978-0-521-87761-9 (hardback : alk. paper) – ISBN 978-0-521-70147-1
(pbk. : alk. paper)
1. Maxwell equations. I. Title.

QC670.F56 2007
530.1401–dc22
2007037901

ISBN 978-0-521-87761-9 Hardback
ISBN 978-0-521-70147-1 Paperback

Contents

Preface

This book has one purpose: to help you understand four of the most influential equations in all of science. If you need a testament to the power of Maxwell's Equations, look around you – radio, television, radar, wireless Internet access, and Bluetooth technology are a few examples of contemporary technology rooted in electromagnetic field theory. Little wonder that the readers of *Physics World* selected Maxwell's Equations as "the most important equations of all time."

How is this book different from the dozens of other texts on electricity and magnetism? Most importantly, the focus is exclusively on Maxwell's Equations, which means you won't have to wade through hundreds of pages of related topics to get to the essential concepts. This leaves room for in-depth explanations of the most relevant features, such as the difference between charge-based and induced electric fields, the physical meaning of divergence and curl, and the usefulness of both the integral and differential forms of each equation.

You'll also find the presentation to be very different from that of other books. Each chapter begins with an "expanded view" of one of Maxwell's Equations, in which the meaning of each term is clearly called out. If you've already studied Maxwell's Equations and you're just looking for a quick review, these expanded views may be all you need. But if you're a bit unclear on any aspect of Maxwell's Equations, you'll find a detailed explanation of every symbol (including the mathematical operators) in the sections following each expanded view. So if you're not sure of the meaning of $\vec{E} \circ \hat{n}$ in Gauss's Law or why it is only the enclosed currents that contribute to the circulation of the magnetic field, you'll want to read those sections.

As a student's guide, this book comes with two additional resources designed to help you understand and apply Maxwell's Equations: an interactive website and a series of audio podcasts. On the website, you'll find the complete solution to every problem presented in the text in interactive format – which means that you'll be able to view the entire solution at once, or ask for a series of helpful hints that will guide you to the final answer. And if you're the kind of learner who benefits from hearing spoken words rather than just reading text, the audio podcasts are for you. These MP3 files walk you through each chapter of the book, pointing out important details and providing further explanations of key concepts.

Is this book right for you? It is if you're a science or engineering student who has encountered Maxwell's Equations in one of your textbooks, but you're unsure of exactly what they mean or how to use them. In that case, you should read the book, listen to the accompanying podcasts, and work through the examples and problems before taking a standardized test such as the Graduate Record Exam. Alternatively, if you're a graduate student reviewing for your comprehensive exams, this book and the supplemental materials will help you prepare.

And if you're neither an undergraduate nor a graduate science student, but a curious young person or a lifelong learner who wants to know more about electric and magnetic fields, this book will introduce you to the four equations that are the basis for much of the technology you use every day.

The explanations in this book are written in an informal style in which mathematical rigor is maintained only insofar as it doesn't get in the way of understanding the physics behind Maxwell's Equations. You'll find plenty of physical analogies – for example, comparison of the flux of electric and magnetic fields to the flow of a physical fluid. James Clerk Maxwell was especially keen on this way of thinking, and he was careful to point out that analogies are useful not because the *quantities* are alike but because of the corresponding *relationships between quantities*. So although nothing is actually flowing in a static electric field, you're likely to find the analogy between a faucet (as a source of fluid flow) and positive electric charge (as the source of electric field lines) very helpful in understanding the nature of the electrostatic field.

One final note about the four Maxwell's Equations presented in this book: it may surprise you to learn that when Maxwell worked out his theory of electromagnetism, he ended up with not four but *twenty* equations that describe the behavior of electric and magnetic fields. It was Oliver Heaviside in Great Britain and Heinrich Hertz in Germany who combined and simplified Maxwell's Equations into four equations in the two decades after Maxwell's death. Today we call these four equations Gauss's law for electric fields, Gauss's law for magnetic fields, Faraday's law, and the Ampere–Maxwell law. Since these four laws are now widely defined as Maxwell's Equations, they are the ones you'll find explained in the book.

Acknowledgments

This book is the result of a conversation with the great Ohio State radio astronomer John Kraus, who taught me the value of plain explanations. Professor Bill Dollhopf of Wittenberg University provided helpful suggestions on the Ampere–Maxwell law, and postdoc Casey Miller of the University of Texas did the same for Gauss's law. The entire manuscript was reviewed by UC Berkeley graduate student Julia Kregenow and Wittenberg undergraduate Carissa Reynolds, both of whom made significant contributions to the content as well as the style of this work. Daniel Gianola of Johns Hopkins University and Wittenberg graduate Melanie Runkel helped with the artwork. The Maxwell Foundation of Edinburgh gave me a place to work in the early stages of this project, and Cambridge University made available their extensive collection of James Clerk Maxwell's papers. Throughout the development process, Dr. John Fowler of Cambridge University Press has provided deft guidance and patient support. And speaking of patience, the amazing Jill Gianola surely holds the modern-day record.

1

Gauss's law for electric fields

In Maxwell's Equations, you'll encounter two kinds of electric field: the *electrostatic* field produced by electric charge and the *induced* electric field produced by a changing magnetic field. Gauss's law for electric fields deals with the electrostatic field, and you'll find this law to be a powerful tool because it relates the spatial behavior of the electrostatic field to the charge distribution that produces it.

1.1 The integral form of Gauss's law

There are many ways to express Gauss's law, and although notation differs among textbooks, the integral form is generally written like this:

$$\oint_S \vec{E} \circ \hat{n}\, da = \frac{q_{enc}}{\varepsilon_0} \quad \text{Gauss's law for electric fields (integral form)}.$$

The left side of this equation is no more than a mathematical description of the electric flux – the number of electric field lines – passing through a closed surface S, whereas the right side is the total amount of charge contained within that surface divided by a constant called the permittivity of free space.

If you're not sure of the exact meaning of "field line" or "electric flux," don't worry – you can read about these concepts in detail later in this chapter. You'll also find several examples showing you how to use Gauss's law to solve problems involving the electrostatic field. For starters, make sure you grasp the main idea of Gauss's law:

> Electric charge produces an electric field, and the flux of that field passing through any closed surface is proportional to the total charge contained within that surface.

In other words, if you have a real or imaginary closed surface of any size and shape and there is no charge inside the surface, the electric flux through the surface must be zero. If you were to place some positive charge anywhere inside the surface, the electric flux through the surface would be positive. If you then added an equal amount of negative charge inside the surface (making the total enclosed charge zero), the flux would again be zero. Remember that it is the *net* charge enclosed by the surface that matters in Gauss's law.

To help you understand the meaning of each symbol in the integral form of Gauss's law for electric fields, here's an expanded view:

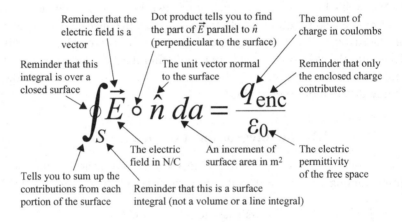

How is Gauss's law useful? There are two basic types of problems that you can solve using this equation:

(1) Given information about a distribution of electric charge, you can find the electric flux through a surface enclosing that charge.
(2) Given information about the electric flux through a closed surface, you can find the total electric charge enclosed by that surface.

The best thing about Gauss's law is that for certain highly symmetric distributions of charges, you can use it to find the electric field itself, rather than just the electric flux over a surface.

Although the integral form of Gauss's law may look complicated, it is completely understandable if you consider the terms one at a time. That's exactly what you'll find in the following sections, starting with $\vec{E}$, the electric field.

$\boxed{\vec{E}}$ The electric field

To understand Gauss's law, you first have to understand the concept of the electric field. In some physics and engineering books, no direct definition of the electric field is given; instead you'll find a statement that an electric field is "said to exist" in any region in which electrical forces act. But what exactly *is* an electric field?

This question has deep philosophical significance, but it is not easy to answer. It was Michael Faraday who first referred to an electric "field of force," and James Clerk Maxwell identified that field as the space around an electrified object – a space in which electric forces act.

The common thread running through most attempts to define the electric field is that fields and forces are closely related. So here's a very pragmatic definition: an electric field is the electrical force per unit charge exerted on a charged object. Although philosophers debate the true meaning of the electric field, you can solve many practical problems by thinking of the electric field at any location as the number of newtons of electrical force exerted on each coulomb of charge at that location. Thus, the electric field may be defined by the relation

$$\vec{E} \equiv \frac{\vec{F}_e}{q_0}, \qquad (1.1)$$

where $\vec{F}_e$ is the electrical force on a small[1] charge q_0. This definition makes clear two important characteristics of the electric field:

(1) $\vec{E}$ is a vector quantity with magnitude directly proportional to force and with direction given by the direction of the force on a positive test charge.
(2) $\vec{E}$ has units of newtons per coulomb (N/C), which are the same as volts per meter (V/m), since volts = newtons × meters/coulombs.

In applying Gauss's law, it is often helpful to be able to visualize the electric field in the vicinity of a charged object. The most common approaches to constructing a visual representation of an electric field are to use either arrows or "field lines" that point in the direction of the field at each point in space. In the arrow approach, the strength of the field is indicated by the length of the arrow, whereas in the field line

---

[1] Why do physicists and engineers always talk about small test charges? Because the job of this charge is to *test* the electric field at a location, not to add another electric field into the mix (although you can't stop it from doing so). Making the test charge infinitesimally small minimizes the effect of the test charge's own field.

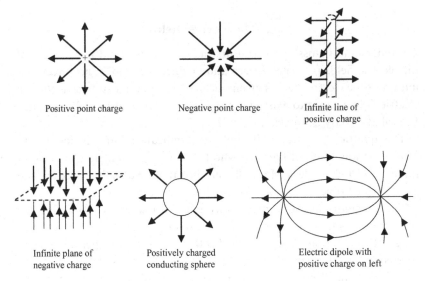

Figure 1.1 Examples of electric fields. Remember that these fields exist in three dimensions; full three-dimensional (3-D) visualizations are available on the book's website.

approach, it is the spacing of the lines that tells you the field strength (with closer lines signifying a stronger field). When you look at a drawing of electric field lines or arrows, be sure to remember that the field exists between the lines as well.

Examples of several electric fields relevant to the application of Gauss's law are shown in Figure 1.1.

Here are a few rules of thumb that will help you visualize and sketch the electric fields produced by charges[2]:

- Electric field lines must originate on positive charge and terminate on negative charge.
- The net electric field at any point is the vector sum of all electric fields present at that point.
- Electric field lines can never cross, since that would indicate that the field points in two different directions at the same location (if two or more different sources contribute electric fields pointing in different directions at the same location, the total electric field is the vector sum

---

[2] In Chapter 3, you can read about electric fields produced not by charges but by changing magnetic fields. That type of field circulates back on itself and does not obey the same rules as electric fields produced by charge.

Table 1.1. *Electric field equations for simple objects*

Point charge (charge $= q$)	$\vec{E} = \dfrac{1}{4\pi\varepsilon_0}\dfrac{q}{r^2}\hat{r}$ (at distance r from q)
Conducting sphere (charge $= Q$)	$\vec{E} = \dfrac{1}{4\pi\varepsilon_0}\dfrac{Q}{r^2}\hat{r}$ (outside, distance r from center) $\vec{E} = 0$ (inside)
Uniformly charged insulating sphere (charge $= Q$, radius $= r_0$)	$\vec{E} = \dfrac{1}{4\pi\varepsilon_0}\dfrac{Q}{r^2}\hat{r}$ (outside, distance r from center) $\vec{E} = \dfrac{1}{4\pi\varepsilon_0}\dfrac{Qr}{r_0^3}\hat{r}$ (inside, distance r from center)
Infinite line charge (linear charge density $= \lambda$)	$\vec{E} = \dfrac{1}{2\pi\varepsilon_0}\dfrac{\lambda}{r}\hat{r}$ (distance r from line)
Infinite flat plane (surface charge density $= \sigma$)	$\vec{E} = \dfrac{\sigma}{2\varepsilon_0}\hat{n}$

of the individual fields, and the electric field lines always point in the single direction of the total field).

- Electric field lines are always perpendicular to the surface of a conductor in equilibrium.

Equations for the electric field in the vicinity of some simple objects may be found in Table 1.1.

So exactly what does the $\vec{E}$ in Gauss's law represent? It represents the total electric field at each point on the surface under consideration. The surface may be real or imaginary, as you'll see when you read about the meaning of the surface integral in Gauss's law. But first you should consider the dot product and unit normal that appear inside the integral.

⊡ The dot product

When you're dealing with an equation that contains a multiplication symbol (a circle or a cross), it is a good idea to examine the terms on both sides of that symbol. If they're printed in bold font or are wearing vector hats (as are $\vec{E}$ and $\hat{n}$ in Gauss's law), the equation involves vector multiplication, and there are several different ways to multiply vectors (quantities that have both magnitude and direction).

In Gauss's law, the circle between $\vec{E}$ and $\hat{n}$ represents the dot product (or "scalar product") between the electric field vector $\vec{E}$ and the unit normal vector $\hat{n}$ (discussed in the next section). If you know the Cartesian components of each vector, you can compute this as

$$\vec{A} \circ \vec{B} = A_x B_x + A_y B_y + A_z B_z. \tag{1.2}$$

Or, if you know the angle θ between the vectors, you can use

$$\vec{A} \circ \vec{B} = |\vec{A}||\vec{B}| \cos \theta, \tag{1.3}$$

where $|\vec{A}|$ and $|\vec{B}|$ represent the magnitude (length) of the vectors. Notice that the dot product between two vectors gives a *scalar* result.

To grasp the physical significance of the dot product, consider vectors $\vec{A}$ and $\vec{B}$ that differ in direction by angle θ, as shown in Figure 1.2(a).

For these vectors, the projection of $\vec{A}$ onto $\vec{B}$ is $|\vec{A}| \cos \theta$, as shown in Figure 1.2(b). Multiplying this projection by the length of $\vec{B}$ gives $|\vec{A}||\vec{B}| \cos \theta$. Thus, the dot product $\vec{A} \circ \vec{B}$ represents the projection of $\vec{A}$ onto the direction of $\vec{B}$ multiplied by the length of $\vec{B}$.[3] The usefulness of this operation in Gauss's law will become clear once you understand the meaning of the vector $\hat{n}$.

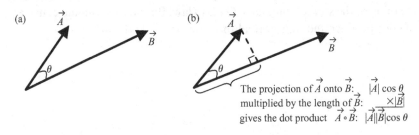

Figure 1.2 The meaning of the dot product.

___

[3] You could have obtained the same result by finding the projection of $\vec{B}$ onto the direction of $\vec{A}$ and then multiplying by the length of $\vec{A}$.

$\boxed{\hat{n}}$ The unit normal vector

The concept of the unit normal vector is straightforward; at any point on a surface, imagine a vector with length of one pointing in the direction perpendicular to the surface. Such a vector, labeled $\hat{n}$, is called a "unit" vector because its length is unity and "normal" because it is perpendicular to the surface. The unit normal for a planar surface is shown in Figure 1.3(a).

Certainly, you could have chosen the unit vector for the plane in Figure 1.3(a) to point in the opposite direction – there's no fundamental difference between one side of an open surface and the other (recall that an open surface is any surface for which it is possible to get from one side to the other without going *through* the surface).

For a closed surface (defined as a surface that divides space into an "inside" and an "outside"), the ambiguity in the direction of the unit normal has been resolved. By convention, the unit normal vector for a closed surface is taken to point outward – away from the volume enclosed by the surface. Some of the unit vectors for a sphere are shown in Figure 1.3(b); notice that the unit normal vectors at the Earth's North and South Pole would point in opposite directions if the Earth were a perfect sphere.

You should be aware that some authors use the notation $d\vec{a}$ rather than $\hat{n}\,da$. In that notation, the unit normal is incorporated into the vector area element $d\vec{a}$, which has magnitude equal to the area da and direction along the surface normal $\hat{n}$. Thus $d\vec{a}$ and $\hat{n}\,da$ serve the same purpose.

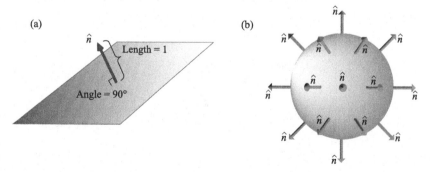

Figure 1.3 Unit normal vectors for planar and spherical surfaces.

$\boxed{\vec{E} \circ \hat{n}}$ The component of $\vec{E}$ normal to a surface

If you understand the dot product and unit normal vector, the meaning of $\vec{E} \circ \hat{n}$ should be clear; this expression represents the component of the electric field vector that is perpendicular to the surface under consideration.

If the reasoning behind this statement isn't apparent to you, recall that the dot product between two vectors such as $\vec{E}$ and $\hat{n}$ is simply the projection of the first onto the second multiplied by the length of the second. Recall also that by definition the length of the unit normal is one ($|\hat{n}| = 1$), so that

$$\vec{E} \circ \hat{n} = |\vec{E}||\hat{n}| \cos\theta = |\vec{E}| \cos\theta, \tag{1.4}$$

where θ is the angle between the unit normal $\hat{n}$ and $\vec{E}$. This is the component of the electric field vector perpendicular to the surface, as illustrated in Figure 1.4.

Thus, if $\theta = 90°$, $\vec{E}$ is perpendicular to $\hat{n}$, which means that the electric field is parallel to the surface, and $\vec{E} \circ \hat{n} = |\vec{E}| \cos(90°) = 0$. So in this case the component of $\vec{E}$ perpendicular to the surface is zero.

Conversely, if $\theta = 0°$, $\vec{E}$ is parallel to $\hat{n}$, meaning the electric field is perpendicular to the surface, and $\vec{E} \circ \hat{n} = |\vec{E}| \cos(0°) = |\vec{E}|$. In this case, the component of $\vec{E}$ perpendicular to the surface is the entire length of $\vec{E}$.

The importance of the electric field component normal to the surface will become clear when you consider electric flux. To do that, you should make sure you understand the meaning of the surface integral in Gauss's law.

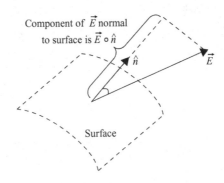

Figure 1.4 Projection of $\vec{E}$ onto direction of $\hat{n}$.

$\int_S () \, da$ The surface integral

Many equations in physics and engineering – Gauss's law among them – involve the area integral of a scalar function or vector field over a specified surface (this type of integral is also called the "surface integral"). The time you spend understanding this important mathematical operation will be repaid many times over when you work problems in mechanics, fluid dynamics, and electricity and magnetism (E&M).

The meaning of the surface integral can be understood by considering a thin surface such as that shown in Figure 1.5. Imagine that the area density (the mass per unit area) of this surface varies with x and y, and you want to determine the total mass of the surface. You can do this by dividing the surface into two-dimensional segments over each of which the area density is approximately constant.

For individual segments with area density σ_i and area dA_i, the mass of each segment is $\sigma_i \, dA_i$, and the mass of the entire surface of N segments is given by $\sum_{i=1}^{N} \sigma_i \, dA_i$. As you can imagine, the smaller you make the area segments, the closer this gets to the true mass, since your approximation of constant σ is more accurate for smaller segments. If you let the segment area dA approach zero and N approach infinity, the summation becomes integration, and you have

$$\text{Mass} = \int_S \sigma(x, y) \, dA.$$

This is the area integral of the scalar function $\sigma(x, y)$ over the surface S. It is simply a way of adding up the contributions of little pieces of a function (the density in this case) to find a total quantity. To understand the integral form of Gauss's law, it is necessary to extend the concept of the surface integral to vector fields, and that's the subject of the next section.

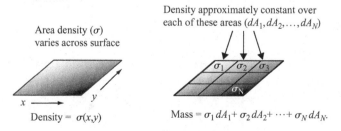

Figure 1.5 Finding the mass of a variable-density surface.

$$\boxed{\int_S \vec{A} \circ \hat{n}\, da}\ \text{The flux of a vector field}$$

In Gauss's law, the surface integral is applied not to a scalar function (such as the density of a surface) but to a vector field. What's a vector field? As the name suggests, a vector field is a distribution of quantities in space – a field – and these quantities have both magnitude and direction, meaning that they are vectors. So whereas the distribution of temperature in a room is an example of a scalar field, the speed and direction of the flow of a fluid at each point in a stream is an example of a vector field.

The analogy of fluid flow is very helpful in understanding the meaning of the "flux" of a vector field, even when the vector field is static and nothing is actually flowing. You can think of the flux of a vector field over a surface as the "amount" of that field that "flows" through that surface, as illustrated in Figure 1.6.

In the simplest case of a uniform vector field $\vec{A}$ and a surface S perpendicular to the direction of the field, the flux Φ is defined as the product of the field magnitude and the area of the surface:

$$\Phi = |\vec{A}| \times \text{surface area}. \tag{1.5}$$

This case is shown in Figure 1.6(a). Note that if $\vec{A}$ is perpendicular to the surface, it is parallel to the unit normal $\hat{n}$.

If the vector field is uniform but is not perpendicular to the surface, as in Figure 1.6(b), the flux may be determined simply by finding the component of $\vec{A}$ perpendicular to the surface and then multiplying that value by the surface area:

$$\Phi = \vec{A} \circ \hat{n} \times (\text{surface area}). \tag{1.6}$$

While uniform fields and flat surfaces are helpful in understanding the concept of flux, many E&M problems involve nonuniform fields and curved surfaces. To work those kinds of problems, you'll need to understand how to extend the concept of the surface integral to vector fields.

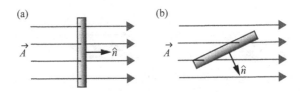

Figure 1.6 Flux of a vector field through a surface.

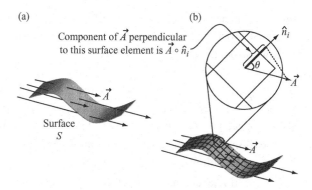

Figure 1.7 Component of $\vec{A}$ perpendicular to surface.

Consider the curved surface and vector field $\vec{A}$ shown in Figure 1.7(a). Imagine that $\vec{A}$ represents the flow of a real fluid and S a porous membrane; later you'll see how this applies to the flux of an electric field through a surface that may be real or purely imaginary.

Before proceeding, you should think for a moment about how you might go about finding the rate of flow of material through surface S. You can define "rate of flow" in a few different ways, but it will help to frame the question as "How many particles pass through the membrane each second?"

To answer this question, define $\vec{A}$ as the number density of the fluid (particles per cubic meter) times the velocity of the flow (meters per second). As the product of the number density (a scalar) and the velocity (a vector), $\vec{A}$ must be a vector in the same direction as the velocity, with units of particles per square meter per second. Since you're trying to find the number of particles per second passing through the surface, dimensional analysis suggests that you multiply $\vec{A}$ by the area of the surface.

But look again at Figure 1.7(a). The different lengths of the arrows are meant to suggest that the flow of material is not spatially uniform, meaning that the speed may be higher or lower at various locations within the flow. This fact alone would mean that material flows through some portions of the surface at a higher rate than other portions, but you must also consider the *angle* of the surface to the direction of flow. Any portion of the surface lying precisely along the direction of flow will necessarily have zero particles per second passing through it, since the flow lines must penetrate the surface to carry particles from one side to the other. Thus, you must be concerned not only with the speed of flow and the area of each portion of the membrane, but also with the component of the flow perpendicular to the surface.

Of course, you know how to find the component of $\vec{A}$ perpendicular to the surface; simply form the dot product of $\vec{A}$ and $\hat{n}$, the unit normal to the surface. But since the surface is curved, the direction of $\hat{n}$ depends on which part of the surface you're considering. To deal with the different $\hat{n}$ (and $\vec{A}$) at each location, divide the surface into small segments, as shown in Figure 1.7(b). If you make these segments sufficiently small, you can assume that both $\hat{n}$ and $\vec{A}$ are constant over each segment.

Let $\hat{n}_i$ represent the unit normal for the ith segment (of area da_i); the flow through segment i is $(\vec{A}_i \circ \hat{n}_i) \, da_i$, and the total is

$$\text{flow through entire surface} = \sum_i \vec{A}_i \circ \hat{n}_i \, da_i.$$

It should come as no surprise that if you now let the size of each segment shrink to zero, the summation becomes integration.

$$\text{Flow through entire surface} = \int_S \vec{A} \circ \hat{n} \, da. \tag{1.7}$$

For a closed surface, the integral sign includes a circle:

$$\oint_S \vec{A} \circ \hat{n} \, da. \tag{1.8}$$

This flow is the particle flux through a closed surface S, and the similarity to the left side of Gauss's law is striking. You have only to replace the vector field $\vec{A}$ with the electric field $\vec{E}$ to make the expressions identical.

$\boxed{\oint_S \vec{E} \circ \hat{n} \, da}$ **The electric flux through a closed surface**

On the basis of the results of the previous section, you should understand that the flux Φ_E of vector field $\vec{E}$ through surface S can be determined using the following equations:

$$\Phi_E = |\vec{E}| \times (\text{surface area}) \quad \vec{E} \text{ is uniform and perpendicular to } S, \quad (1.9)$$

$$\Phi_E = \vec{E} \circ \hat{n} \times (\text{surface area}) \quad \vec{E} \text{ is uniform and at an angle to } S, \quad (1.10)$$

$$\Phi_E = \int_S \vec{E} \circ \hat{n} \, da \quad \vec{E} \text{ is non-uniform and at a variable angle to } S. \quad (1.11)$$

These relations indicate that electric flux is a scalar quantity and has units of electric field times area, or Vm. But does the analogy used in the previous section mean that the electric flux should be thought of as a flow of particles, and that the electric field is the product of a density and a velocity?

The answer to this question is "absolutely not." Remember that when you employ a physical analogy, you're hoping to learn something about the *relationships between quantities*, not about the quantities themselves. So, you can find the electric flux by integrating the normal component of the electric field over a surface, but you should not think of the electric flux as the physical movement of particles.

How should you think of electric flux? One helpful approach follows directly from the use of field lines to represent the electric field. Recall that in such representations the strength of the electric field at any point is indicated by the spacing of the field lines at that location. More specifically, the electric field strength can be considered to be proportional to the density of field lines (the number of field lines per square meter) in a plane perpendicular to the field at the point under consideration. Integrating that density over the entire surface gives the number of field lines penetrating the surface, and that is exactly what the expression for electric flux gives. Thus, another way to define electric flux is

electric flux $(\Phi_E) \equiv$ number of field lines penetrating surface.

There are two caveats you should keep in mind when you think of electric flux as the number of electric field lines penetrating a surface. The first is that field lines are only a convenient representation of the electric field, which is actually continuous in space. The number of field lines you

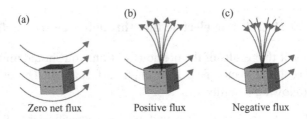

Figure 1.8 Flux lines penetrating closed surfaces.

choose to draw for a given field is up to you, so long as you maintain consistency between fields of different strengths – which means that fields that are twice as strong must be represented by twice as many field lines per unit area.

The second caveat is that surface penetration is a two-way street; once the direction of a surface normal $\hat{n}$ has been established, field line components parallel to that direction give a positive flux, while components in the opposite direction (antiparallel to $\hat{n}$) give a negative flux. Thus, a surface penetrated by five field lines in one direction (say from the top side to the bottom side) and five field lines in the opposite direction (from bottom to top) has zero flux, because the contributions from the two groups of field lines cancel. So, you should think of electric flux as the *net* number of field lines penetrating the surface, with direction of penetration taken into account.

If you give some thought to this last point, you may come to an important conclusion about closed surfaces. Consider the three boxes shown in Figure 1.8. The box in Figure 1.8(a) is penetrated only by electric field lines that originate and terminate outside the box. Thus, every field line that enters must leave, and the flux through the box must be zero.

Remembering that the unit normal for closed surfaces points away from the enclosed volume, you can see that the inward flux (lines entering the box) is negative, since $\vec{E} \circ \hat{n}$ must be negative when the angle between $\vec{E}$ and $\hat{n}$ is greater than 90°. This is precisely cancelled by the outward flux (lines exiting the box), which is positive, since $\vec{E} \circ \hat{n}$ is positive when the angle between $\vec{E}$ and $\hat{n}$ is less than 90°.

Now consider the box in Figure 1.8(b). The surfaces of this box are penetrated not only by the field lines originating outside the box, but also by a group of field lines that originate within the box. In this case, the net number of field lines is clearly not zero, since the positive flux of the lines that originate in the box is not compensated by any incoming (negative) flux. Thus, you can say with certainty that if the flux through any closed surface is positive, that surface must contain a *source* of field lines.

Finally, consider the box in Figure 1.8(c). In this case, some of the field lines terminate within the box. These lines provide a negative flux at the surface through which they enter, and since they don't exit the box, their contribution to the net flux is not compensated by any positive flux. Clearly, if the flux through a closed surface is negative, that surface must contain a *sink* of field lines (sometimes referred to as a drain).

Now recall the first rule of thumb for drawing charge-induced electric field lines; they must originate on positive charge and terminate on negative charge. So, the point from which the field lines diverge in Figure 1.8(b) marks the location of some amount of positive charge, and the point to which the field lines converge in Figure 1.8(c) indicates the existence of negative charge at that location.

If the amount of charge at these locations were greater, there would be more field lines beginning or ending on these points, and the flux through the surface would be greater. And if there were equal amounts of positive and negative charge within one of these boxes, the positive (outward) flux produced by the positive charge would exactly cancel the negative (inward) flux produced by the negative charge. So, in this case the flux would be zero, just as the net charge contained within the box would be zero.

You should now see the physical reasoning behind Gauss's law: the electric flux passing through any closed surface – that is, the number of electric field lines penetrating that surface – must be proportional to the total charge contained within that surface. Before putting this concept to use, you should take a look at the right side of Gauss's law.

$\boxed{q_{enc}}$ The enclosed charge

If you understand the concept of flux as described in the previous section, it should be clear why the right side of Gauss's law involves only the *enclosed* charge – that is, the charge within the closed surface over which the flux is determined. Simply put, it is because any charge located outside the surface produces an equal amount of inward (negative) flux and outward (positive) flux, so the net contribution to the flux through the surface must be zero.

How can you determine the charge enclosed by a surface? In some problems, you're free to choose a surface that surrounds a known amount of charge, as in the situations shown in Figure 1.9. In each of these cases, the total charge within the selected surface can be easily determined from geometric considerations.

For problems involving groups of discrete charges enclosed by surfaces of any shape, finding the total charge is simply a matter of adding the individual charges.

$$\text{Total enclosed charge} = \sum_i q_i.$$

While small numbers of discrete charges may appear in physics and engineering problems, in the real world you're far more likely to encounter charged objects containing billions of charge carriers lined along a wire, slathered over a surface, or arrayed throughout a volume. In such cases, counting the individual charges is not practical – but you can determine the total charge if you know the charge density. Charge density may be specified in one, two, or three dimensions (1-, 2-, or 3-D).

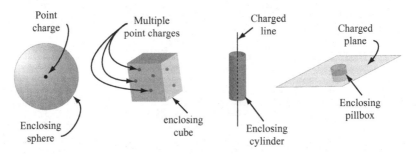

Figure 1.9 Surface enclosing known charges.

Dimensions	Terminology	Symbol	Units
1	Linear charge density	λ	C/m
2	Area charge density	σ	C/m^2
3	Volume charge density	ρ	C/m^3

If these quantities are constant over the length, area, or volume under consideration, finding the enclosed charge requires only a single multiplication:

$$\text{1-D}: \quad q_{\text{enc}} = \lambda L \quad (L = \text{enclosed length of charged line}), \qquad (1.12)$$

$$\text{2-D}: \quad q_{\text{enc}} = \sigma A \quad (A = \text{enclosed area of charged surface}), \qquad (1.13)$$

$$\text{3-D}: \quad q_{\text{enc}} = \rho V \quad (V = \text{enclosed portion of charged volume}). \qquad (1.14)$$

You are also likely to encounter situations in which the charge density is not constant over the line, surface, or volume of interest. In such cases, the integration techniques described in the "Surface Integral" section of this chapter must be used. Thus,

$$\text{1-D}: \quad q_{\text{enc}} = \int_L \lambda \, dl \text{ where } \lambda \text{ varies along a line,} \qquad (1.15)$$

$$\text{2-D}: \quad q_{\text{enc}} = \int_S \sigma \, da \text{ where } \sigma \text{ varies over a surface,} \qquad (1.16)$$

$$\text{3-D}: \quad q_{\text{enc}} = \int_V \rho \, dV \text{ where } \rho \text{ varies over a volume.} \qquad (1.17)$$

You should note that the enclosed charge in Gauss's law for electric fields is the *total* charge, including both free and bound charge. You can read about bound charge in the next section, and you'll find a version of Gauss's law that depends only on free charge in the Appendix.

Once you've determined the charge enclosed by a surface of any size and shape, it is very easy to find the flux through that surface; simply divide the enclosed charge by ε_0, the permittivity of free space. The physical meaning of that parameter is described in the next section.

$\boxed{\varepsilon_0}$ The permittivity of free space

The constant of proportionality between the electric flux on the left side of Gauss's law and the enclosed charge on the right side is ε_0, the permittivity of free space. The permittivity of a material determines its response to an applied electric field – in nonconducting materials (called "insulators" or "dielectrics"), charges do not move freely, but may be slightly displaced from their equilibrium positions. The relevant permittivity in Gauss's law for electric fields is the permittivity of free space (or "vacuum permittivity"), which is why it carries the subscript zero.

The value of the vacuum permittivity in SI units is approximately 8.85×10^{-12} coulombs per volt-meter (C/Vm); you will sometimes see the units of permittivity given as farads per meter (F/m), or, more fundamentally, $(C^2s^2/kg\ m^3)$. A more precise value for the permittivity of free space is

$$\varepsilon_0 = 8.8541878176 \times 10^{-12} \text{ C/Vm}.$$

Does the presence of this quantity mean that this form of Gauss's law is only valid in a vacuum? No, Gauss's law as written in this chapter is general, and applies to electric fields within dielectrics as well as those in free space, provided that you account for *all* of the enclosed charge, including charges that are bound to the atoms of the material.

The effect of bound charges can be understood by considering what happens when a dielectric is placed in an external electric field. Inside the dielectric material, the amplitude of the total electric field is generally less than the amplitude of the applied field.

The reason for this is that dielectrics become "polarized" when placed in an electric field, which means that positive and negative charges are displaced from their original positions. And since positive charges are displaced in one direction (parallel to the applied electric field) and negative charges are displaced in the opposite direction (antiparallel to the applied field), these displaced charges give rise to their own electric field that opposes the external field, as shown in Figure 1.10. This makes the net field within the dielectric less than the external field.

It is the ability of dielectric materials to reduce the amplitude of an electric field that leads to their most common application: increasing the capacitance and maximum operating voltage of capacitors. As you may recall, the capacitance (ability to store charge) of a parallel-plate capacitor is

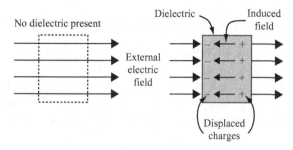

Figure 1.10 Electric field induced in a dielectric.

$$C = \frac{\varepsilon A}{d},$$

where A is the plate area, d is the plate separation, and ε is the permittivity of the material between the plates. High-permittivity materials can provide increased capacitance without requiring larger plate area or decreased plate spacing.

The permittivity of a dielectric is often expressed as the relative permittivity, which is the factor by which the material's permittivity exceeds that of free space:

$$\text{relative permittivity } \varepsilon_r = \varepsilon/\varepsilon_0.$$

Some texts refer to relative permittivity as "dielectric constant," although the variation in permittivity with frequency suggests that the word "constant" is better used elsewhere. The relative permittivity of ice, for example, changes from approximately 81 at frequencies below 1 kHz to less than 5 at frequencies above 1 MHz. Most often, it is the low-frequency value of permittivity that is called the dielectric constant.

One more note about permittivity; as you'll see in Chapter 5, the permittivity of a medium is a fundamental parameter in determining the speed with which an electromagnetic wave propagates through that medium.

$\boxed{\oint_s \vec{E} \circ \hat{n} \, da = q_{\text{enc}}/\varepsilon_0}$ **Applying Gauss's law (integral form)**

A good test of your understanding of an equation like Gauss's law is whether you're able to solve problems by applying it to relevant situations. At this point, you should be convinced that Gauss's law relates the electric flux through a closed surface to the charge enclosed by that surface. Here are some examples of what you can actually *do* with that information.

Example 1.1: Given a charge distribution, find the flux through a closed surface surrounding that charge.

Problem: Five point charges are enclosed in a cylindrical surface S. If the values of the charges are $q_1 = +3 \, \text{nC}$, $q_2 = -2 \, \text{nC}$, $q_3 = +2 \, \text{nC}$, $q_4 = +4 \, \text{nC}$, and $q_5 = -1 \, \text{nC}$, find the total flux through S.

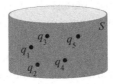

Solution: From Gauss's law,

$$\Phi_E = \oint_S \vec{E} \circ \hat{n} \, da = \frac{q_{\text{enc}}}{\varepsilon_0}.$$

For discrete charges, you know that the total charge is just the sum of the individual charges. Thus,

$$q_{\text{enc}} = \text{Total enclosed charge} = \sum_i q_i$$
$$= (3 - 2 + 2 + 4 - 1) \times 10^{-9} \, \text{C}$$
$$= 6 \times 10^{-9} \, \text{C}$$

and

$$\Phi_E = \frac{q_{\text{enc}}}{\varepsilon_0} = \frac{6 \times 10^{-9} \, \text{C}}{8.85 \times 10^{-12} \, \text{C/Vm}} = 678 \, \text{Vm}.$$

This is the total flux through *any* closed surface surrounding this group of charges.

Example 1.2: Given the flux through a closed surface, find the enclosed charge.

Problem: A line charge with linear charge density $\lambda = 10^{-12}$ C/m passes through the center of a sphere. If the flux through the surface of the sphere is 1.13×10^{-3} Vm, what is the radius R of the sphere?

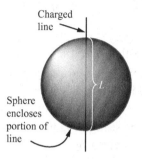

Charged line

Sphere encloses portion of line

L

Solution: The charge on a line charge of length L is given by $q = \lambda L$. Thus,

$$\Phi_E = \frac{q_{enc}}{\varepsilon_0} = \frac{\lambda L}{\varepsilon_0},$$

and

$$L = \frac{\Phi_E \varepsilon_0}{\lambda}.$$

Since L is twice the radius of the sphere, this means

$$2R = \frac{\Phi_E \varepsilon_0}{\lambda} \quad \text{or} \quad R = \frac{\Phi_E \varepsilon_0}{2\lambda}.$$

Inserting the values for Φ_E, ε_0 and λ, you will find that $R = 5 \times 10^{-3}$ m.

Example 1.3: Find the flux through a section of a closed surface.

Problem: A point source of charge q is placed at the center of curvature of a spherical section that extends from spherical angle θ_1 to θ_2 and from ϕ_1 to ϕ_2. Find the electric flux through the spherical section.

Solution: Since the surface of interest in this problem is open, you'll have to find the electric flux by integrating the normal component of the electric field over the surface. You can then check your answer using Gauss's law by allowing the spherical section to form a complete sphere that encloses the point charge.

The electric flux Φ_E is $\int_S \vec{E} \circ \hat{n} \, da$, where S is the spherical section of interest and $\vec{E}$ is the electric field on the surface due to the point charge at the center of curvature, a distance r from the section of interest. From Table 1.1, you know that the electric field at a distance r from a point charge is

$$\vec{E} = \frac{1}{4\pi\varepsilon_0} \frac{q}{r^2} \hat{r}.$$

Before you can integrate this over the surface of interest, you have to consider $\vec{E} \circ \hat{n}$ (that is, you must find the component of the electric field perpendicular to the surface). That is trivial in this case, because the unit normal $\hat{n}$ for a spherical section points in the outward radial direction (the $\hat{r}$ direction), as may be seen in Figure 1.11. This means that $\vec{E}$ and $\hat{n}$ are parallel, and the flux is given by

$$\Phi_E = \int_S \vec{E} \circ \hat{n} \, da = \int_S |\vec{E}||\hat{n}| \cos(0^\circ) \, da = \int_S |\vec{E}| \, da = \int_S \frac{1}{4\pi\varepsilon_0} \frac{q}{r^2} \, da.$$

Since you are integrating over a spherical section in this case, the logical choice for coordinate system is spherical. This makes the area element $r^2 \sin\theta \, d\theta \, d\phi$, and the surface integral becomes

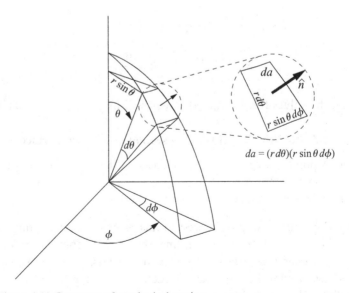

$$da = (r \, d\theta)(r \sin\theta \, d\phi)$$

Figure 1.11 Geometry of a spherical section.

$$\Phi_E = \int_\theta \int_\phi \frac{1}{4\pi\varepsilon_0} \frac{q}{r^2} r^2 \sin\theta \, d\theta \, d\phi = \frac{q}{4\pi\varepsilon_0} \int_\theta \sin\theta \, d\theta \int_\phi d\phi,$$

which is easily integrated to give

$$\Phi_E = \frac{q}{4\pi\varepsilon_0}(\cos\theta_1 - \cos\theta_2)(\phi_2 - \phi_1).$$

As a check on this result, take the entire sphere as the section ($\theta_1 = 0$, $\theta_2 = \pi$, $\phi_1 = 0$, and $\phi_2 = 2\pi$). This gives

$$\Phi_E = \frac{q}{4\pi\varepsilon_0}(1 - (-1))\,(2\pi - 0) = \frac{q}{\varepsilon_0},$$

exactly as predicted by Gauss's law.

Example 1.4: Given $\vec{E}$ over a surface, find the flux through the surface and the charge enclosed by the surface.

Problem: The electric field at distance r from an infinite line charge with linear charge density λ is given in Table 1.1 as

$$\vec{E} = \frac{1}{2\pi\varepsilon_0} \frac{\lambda}{r} \hat{r}.$$

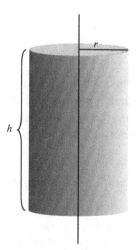

Use this expression to find the electric flux through a cylinder of radius r and height h surrounding a portion of an infinite line charge, and then use Gauss's law to verify that the enclosed charge is λh.

Solution: Problems like this are best approached by considering the flux through each of three surfaces that comprise the cylinder: the top, bottom, and curved side surfaces. The most general expression for the electric flux through any surface is

$$\Phi_E = \int_S \vec{E} \circ \hat{n} \, da,$$

which in this case gives

$$\Phi_E = \int_S \frac{1}{2\pi\varepsilon_0} \frac{\lambda}{r} \hat{r} \circ \hat{n} \, da.$$

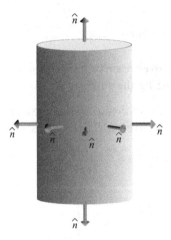

Consider now the unit normal vectors of each of the three surfaces: since the electric field points radially outward from the axis of the cylinder, $\vec{E}$ is perpendicular to the normal vectors of the top and bottom surfaces and parallel to the normal vectors for the curved side of the cylinder. You may therefore write

$$\Phi_{E,\,\text{top}} = \int_S \frac{1}{2\pi\varepsilon_0} \frac{\lambda}{r} \hat{r} \circ \hat{n}_{\text{top}} \, da = 0,$$

$$\Phi_{E,\,\text{bottom}} = \int_S \frac{1}{2\pi\varepsilon_0} \frac{\lambda}{r} \hat{r} \circ \hat{n}_{\text{bottom}} \, da = 0,$$

$$\Phi_{E,\,\text{side}} = \int_S \frac{1}{2\pi\varepsilon_0} \frac{\lambda}{r} \hat{r} \circ \hat{n}_{\text{side}} \, da = \frac{1}{2\pi\varepsilon_0} \frac{\lambda}{r} \int_S da,$$

and, since the area of the curved side of the cylinder is $2\pi rh$, this gives

$$\Phi_{E,\text{side}} = \frac{1}{2\pi\varepsilon_0}\frac{\lambda}{r}(2\pi rh) = \frac{\lambda h}{\varepsilon_0}.$$

Gauss's law tells you that this must equal $q_{\text{enc}}/\varepsilon_0$, which verifies that the enclosed charge $q_{\text{enc}} = \lambda h$ in this case.

Example 1.5: Given a symmetric charge distribution, find $\vec{E}$.

Finding the electric field using Gauss's law may seem to be a hopeless task. After all, while the electric field does appear in the equation, it is only the normal component that emerges from the dot product, and it is only the integral of that normal component over the entire surface that is proportional to the enclosed charge. Do realistic situations exist in which it is possible to dig the electric field out of its interior position in Gauss's law?

Happily, the answer is yes; you may indeed find the electric field using Gauss's law, albeit only in situations characterized by high symmetry. Specifically, you can determine the electric field whenever you're able to design a real or imaginary "special Gaussian surface" that encloses a known amount of charge. A special Gaussian surface is one on which

(1) the electric field is either parallel or perpendicular to the surface normal (which allows you to convert the dot product into an algebraic multiplication), and
(2) the electric field is constant or zero over sections of the surface (which allows you to remove the electric field from the integral).

Of course, the electric field on any surface that you can imagine around arbitrarily shaped charge distributions will not satisfy either of these requirements. But there are situations in which the distribution of charge is sufficiently symmetric that a special Gaussian surface may be imagined. Specifically, the electric field in the vicinity of spherical charge distributions, infinite lines of charge, and infinite planes of charge may be determined by direct application of the integral form of Gauss's law. Geometries that approximate these ideal conditions, or can be approximated by combinations of them, may also be attacked using Gauss's law.

The following problem shows how to use Gauss's law to find the electric field around a spherical distribution of charge; the other cases are covered in the problem set, for which solutions are available on the website.

Problem: Use Gauss's law to find the electric field at a distance r from the center of a sphere with uniform volume charge density ρ and radius a.

Solution: Consider first the electric field outside the sphere. Since the distribution of charge is spherically symmetric, it is reasonable to expect the electric field to be entirely radial (that is, pointed toward or away from the sphere). If that's not obvious to you, imagine what would happen if the electric field had a nonradial component (say in the $\hat{\theta}$ or $\hat{\varphi}$ direction); by rotating the sphere about some arbitrary axis, you'd be able to change the direction of the field. But the charge is uniformly distributed throughout the sphere, so there can be no preferred direction or orientation – rotating the sphere simply replaces one chunk of charge with another, identical chunk – so this can have no effect whatsoever on the electric field. Faced with this conundrum, you are forced to conclude that the electric field of a spherically symmetric charge distribution must be entirely radial.

To find the value of this radial field using Gauss's law, you'll have to imagine a surface that meets the requirements of a special Gaussian surface; $\vec{E}$ must be either parallel or perpendicular to the surface normal at all locations, and $\vec{E}$ must be uniform everywhere on the surface. For a radial electric field, there can be only one choice; your Gaussian surface must be a sphere centered on the charged sphere, as shown in Figure 1.12. Notice that no actual surface need be present, and the special Gaussian surface may be purely imaginary – it is simply a construct that allows you to evaluate the dot product and remove the electric field from the surface integral in Gauss's law.

Since the radial electric field is everywhere parallel to the surface normal, the $\vec{E} \circ \hat{n}$ term in the integral in Gauss's law becomes $|\vec{E}||\hat{n}| \cos(0°)$, and the electric flux over the Gaussian surface S is

$$\Phi_E = \oint_S \vec{E} \circ \hat{n} \, da = \oint_S |\vec{E}| \, da$$

Since $\vec{E}$ has no θ or φ dependence, it must be constant over S, which means it may be removed from the integral:

$$\Phi_E = \oint_S |\vec{E}| \, da = |\vec{E}| \oint_S da = |\vec{E}|(4\pi r^2),$$

where r is the radius of the special Gaussian surface. You can now use Gauss's law to find the value of the electric field:

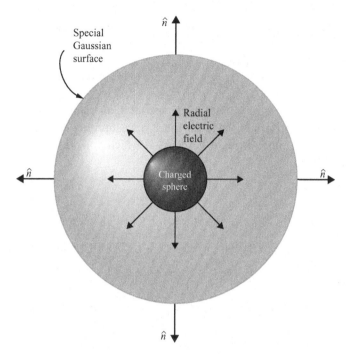

Figure 1.12 A special Gaussian around a charged sphere.

$$\Phi_E = |\vec{E}|(4\pi r^2) = \frac{q_{enc}}{\varepsilon_0},$$

or

$$|\vec{E}| = \frac{q_{enc}}{4\pi\varepsilon_0 r^2},$$

where q_{enc} is the charge enclosed by your Gaussian surface. You can use this expression to find the electric field both outside and inside the sphere.

To find the electric field outside the sphere, construct your Gaussian surface with radius $r > a$ so that the entire charged sphere is within the Gaussian surface. This means that the enclosed charge is just the charge density times the entire volume of the charged sphere: $q_{enc} = (4/3)\pi a^3 \rho$. Thus,

$$|\vec{E}| = \frac{(4/3)\pi a^3 \rho}{4\pi\varepsilon_0 r^2} = \frac{\rho a^3}{3\varepsilon_0 r^2} \quad \text{(outside sphere)}.$$

To find the electric field within the charged sphere, construct your Gaussian surface with $r < a$. In this case, the enclosed charge is the charge density times the volume of your Gaussian surface: $q_{enc} = (4/3)\pi r^3 \rho$. Thus,

$$|\vec{E}| = \frac{(4/3)\pi r^3 \rho}{4\pi\varepsilon_0 r^2} = \frac{\rho r}{3\varepsilon_0} \quad \text{(inside sphere).}$$

The keys to successfully employing special Gaussian surfaces are to recognize the appropriate shape for the surface and then to adjust its size to ensure that it runs through the point at which you wish to determine the electric field.

1.2 The differential form of Gauss's law

The integral form of Gauss's law for electric fields relates the electric flux over a surface to the charge enclosed by that surface – but like all of Maxwell's Equations, Gauss's law may also be cast in *differential* form. The differential form is generally written as

$$\vec{\nabla} \circ \vec{E} = \frac{\rho}{\varepsilon_0} \qquad \text{Gauss's law for electric fields (differential form).}$$

The left side of this equation is a mathematical description of the divergence of the electric field – the tendency of the field to "flow" away from a specified location – and the right side is the electric charge density divided by the permittivity of free space.

Don't be concerned if the del operator ($\vec{\nabla}$) or the concept of divergence isn't perfectly clear to you – these are discussed in the following sections. For now, make sure you grasp the main idea of Gauss's law in differential form:

The electric field produced by electric charge diverges from positive charge and converges upon negative charge.

In other words, the only places at which the divergence of the electric field is not zero are those locations at which charge is present. If positive charge is present, the divergence is positive, meaning that the electric field tends to "flow" away from that location. If negative charge is present, the divergence is negative, and the field lines tend to "flow" toward that point.

Note that there's a fundamental difference between the differential and the integral form of Gauss's law; the differential form deals with the divergence of the electric field and the charge density *at individual points* in space, whereas the integral form entails the integral of the normal component of the electric field *over a surface*. Familiarity with both forms will allow you to use whichever is better suited to the problem you're trying to solve.

To help you understand the meaning of each symbol in the differential form of Gauss's law for electric fields, here's an expanded view:

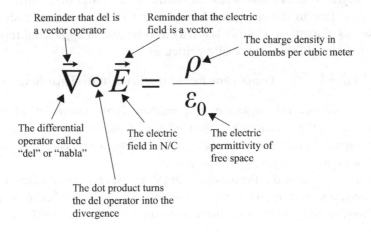

How is the differential form of Gauss's law useful? In any problem in which the spatial variation of the vector electric field is known at a specified location, you can find the volume charge density at that location using this form. And if the volume charge density is known, the divergence of the electric field may be determined.

$\boxed{\vec{\nabla}}$ Nabla – the del operator

An inverted uppercase delta appears in the differential form of all four of Maxwell's Equations. This symbol represents a vector differential operator called "nabla" or "del," and its presence instructs you to take derivatives of the quantity on which the operator is acting. The exact form of those derivatives depends on the symbol following the del operator, with "$\vec{\nabla}\circ$" signifying divergence, "$\vec{\nabla}\times$" indicating curl, and $\vec{\nabla}$ signifying gradient. Each of these operations is discussed in later sections; for now we'll just consider what an operator is and how the del operator can be written in Cartesian coordinates.

Like all good mathematical operators, del is an action waiting to happen. Just as $\sqrt{\ }$ tells you to take the square root of anything that appears under its roof, $\vec{\nabla}$ is an instruction to take derivatives in three directions. Specifically,

$$\vec{\nabla} \equiv \hat{i}\frac{\partial}{\partial x} + \hat{j}\frac{\partial}{\partial y} + \hat{k}\frac{\partial}{\partial z}, \tag{1.18}$$

where $\hat{i}$, $\hat{j}$, and $\hat{k}$ are the unit vectors in the direction of the Cartesian coordinates x, y, and z. This expression may appear strange, since in this form it is lacking anything on which it can operate. In Gauss's law for electric fields, the del operator is dotted into the electric field vector, forming the divergence of $\vec{E}$. That operation and its results are described in the next section.

$\boxed{\vec{\nabla} \circ}$ Del dot – the divergence

The concept of divergence is important in many areas of physics and engineering, especially those concerned with the behavior of vector fields. James Clerk Maxwell coined the term "convergence" to describe the mathematical operation that measures the rate at which electric field lines "flow" toward points of negative electric charge (meaning that positive convergence was associated with negative charge). A few years later, Oliver Heaviside suggested the use of the term "divergence" for the same quantity with the opposite sign. Thus, positive divergence is associated with the "flow" of electric field lines away from positive charge.

Both flux and divergence deal with the "flow" of a vector field, but with an important difference; flux is defined over an area, while divergence applies to individual points. In the case of fluid flow, the divergence at any point is a measure of the tendency of the flow vectors to diverge from that point (that is, to carry more material away from it than is brought toward it). Thus points of positive divergence are *sources* (faucets in situations involving fluid flow, positive electric charge in electrostatics), while points of negative divergence are *sinks* (drains in fluid flow, negative charge in electrostatics).

The mathematical definition of divergence may be understood by considering the flux through an infinitesimal surface surrounding the point of interest. If you were to form the ratio of the flux of a vector field $\vec{A}$ through a surface S to the volume enclosed by that surface as the volume shrinks toward zero, you would have the divergence of $\vec{A}$:

$$\text{div}(\vec{A}) = \vec{\nabla} \circ \vec{A} \equiv \lim_{\Delta V \to 0} \frac{1}{\Delta V} \oint_S \vec{A} \circ \hat{n} \, da. \qquad (1.19)$$

While this expression states the relationship between divergence and flux, it is not particularly useful for finding the divergence of a given vector field. You'll find a more user-friendly mathematical expression for divergence later in this section, but first you should take a look at the vector fields shown in Figure 1.13.

To find the locations of positive divergence in each of these fields, look for points at which the flow vectors either spread out or are larger pointing away from the location and shorter pointing toward it. Some authors suggest that you imagine sprinkling sawdust on flowing water to assess the divergence; if the sawdust is dispersed, you have selected a point of positive divergence, while if it becomes more concentrated, you've picked a location of negative divergence.

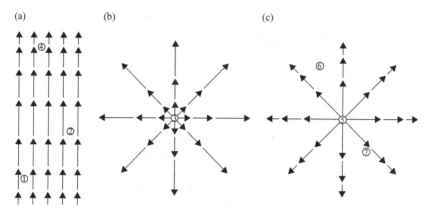

Figure 1.13 Vector fields with various values of divergence.

Using such tests, it is clear that locations such as 1 and 2 in Figure 1.13(a) and location 3 in Figure 1.13(b) are points of positive divergence, while the divergence is negative at point 4.

The divergence at various points in Figure 1.13(c) is less obvious. Location 5 is obviously a point of positive divergence, but what about locations 6 and 7? The flow lines are clearly spreading out at those locations, but they're also getting shorter at greater distance from the center. Does the spreading out compensate for the slowing down of the flow?

Answering that question requires a useful mathematical form of the divergence as well as a description of how the vector field varies from place to place. The differential form of the mathematical operation of divergence or "del dot" ($\vec{\nabla} \circ$) on a vector $\vec{A}$ in Cartesian coordinates is

$$\vec{\nabla} \circ \vec{A} = \left(\hat{i} \frac{\partial}{\partial x} + \hat{j} \frac{\partial}{\partial y} + \hat{k} \frac{\partial}{\partial z} \right) \circ \left(\hat{i} A_x + \hat{j} A_y + \hat{k} A_z \right),$$

and, since $\hat{i} \circ \hat{i} = \hat{j} \circ \hat{j} = \hat{k} \circ \hat{k} = 1$, this is

$$\vec{\nabla} \circ \vec{A} = \left(\frac{\partial A_x}{\partial x} + \frac{\partial A_y}{\partial y} + \frac{\partial A_z}{\partial z} \right). \tag{1.20}$$

Thus, the divergence of the vector field $\vec{A}$ is simply the change in its x-component along the x-axis plus the change in its y-component along the y-axis plus the change in its z-component along the z-axis. Note that the divergence of a vector field is a scalar quantity; it has magnitude but no direction.

You can now apply this to the vector fields in Figure 1.13. In Figure 1.13(a), assume that the magnitude of the vector field varies sinusoidally along the x-axis (which is vertical in this case) as $\vec{A} = \sin(\pi x)\hat{i}$ while remaining constant in the y- and z-directions. Thus,

$$\vec{\nabla} \circ \vec{A} = \frac{\partial A_x}{\partial x} = \pi \cos(\pi x),$$

since A_y and A_z are zero. This expression is positive for $0 < x < \frac{1}{2}$, 0 at $x = \frac{1}{2}$, and negative for $\frac{1}{2} < x < \frac{3}{2}$, just as your visual inspection suggested.

Now consider Figure 1.13(b), which represents a slice through a spherically symmetric vector field with amplitude increasing as the square of the distance from the origin. Thus $\vec{A} = r^2\hat{r}$. Since $r^2 = (x^2 + y^2 + z^2)$ and

$$\hat{r} = \frac{x\hat{i} + y\hat{j} + z\hat{k}}{\sqrt{x^2 + y^2 + z^2}},$$

this means

$$\vec{A} = r^2\hat{r} = (x^2 + y^2 + z^2)\frac{x\hat{i} + y\hat{j} + z\hat{k}}{\sqrt{x^2 + y^2 + z^2}},$$

and

$$\frac{\partial A_x}{\partial x} = (x^2 + y^2 + z^2)^{(1/2)} + x\left(\frac{1}{2}\right)(x^2 + y^2 + z^2)^{-(1/2)}(2x).$$

Doing likewise for the y- and z-components and adding yields

$$\vec{\nabla} \circ \vec{A} = 3(x^2 + y^2 + z^2)^{(1/2)} + \frac{x^2 + y^2 + z^2}{\sqrt{x^2 + y^2 + z^2}} = 4(x^2 + y^2 + z^2)^{1/2} = 4r.$$

Thus, the divergence in the vector field in Figure 1.13(b) is increasing linearly with distance from the origin.

Finally, consider the vector field in Figure 1.13(c), which is similar to the previous case but with the amplitude of the vector field *decreasing* as the square of the distance from the origin. The flow lines are spreading out as they were in Figure 1.13(b), but in this case you might suspect that the decreasing amplitude of the vector field will affect the value of the divergence. Since $\vec{A} = (1/r^2)\hat{r}$,

$$\vec{A} = \frac{1}{(x^2 + y^2 + z^2)}\frac{x\hat{i} + y\hat{j} + z\hat{k}}{\sqrt{x^2 + y^2 + z^2}} = \frac{x\hat{i} + y\hat{j} + z\hat{k}}{(x^2 + y^2 + z^2)^{(3/2)}},$$

and

$$\frac{\partial A_x}{\partial x} = \frac{1}{(x^2 + y^2 + z^2)^{(3/2)}} - x\left(\frac{3}{2}\right)(x^2 + y^2 + z^2)^{-(5/2)}(2x),$$

Adding in the y- and z-derivatives gives

$$\vec{\nabla} \circ \vec{A} = \frac{3}{(x^2 + y^2 + z^2)^{(3/2)}} - \frac{3(x^2 + y^2 + z^2)}{(x^2 + y^2 + z^2)^{(5/2)}} = 0.$$

This validates the suspicion that the reduced amplitude of the vector field with distance from the origin may compensate for the spreading out of the flow lines. Note that this is true only for the case in which the amplitude of the vector field falls off as $1/r^2$ (this case is especially relevant for the electric field, which you'll find in the next section).

As you consider the divergence of the electric field, you should remember that some problems may be solved more easily using non-Cartesian coordinate systems. The divergence may be calculated in cylindrical and spherical coordinate systems using

$$\vec{\nabla} \circ \vec{A} = \frac{1}{r}\frac{\partial}{\partial r}(rA_r) + \frac{1}{r}\frac{\partial A_\phi}{\partial \phi} + \frac{\partial A_z}{\partial z} \quad \text{(cylindrical)}, \tag{1.21}$$

and

$$\vec{\nabla} \circ \vec{A} = \frac{1}{r^2}\frac{\partial}{\partial r}(r^2 A_r) + \frac{1}{r\sin\theta}\frac{\partial}{\partial \theta}(A_\theta \sin\theta)$$
$$+ \frac{1}{r\sin\theta}\frac{\partial A_\phi}{\partial \phi} \quad \text{(spherical)}. \tag{1.22}$$

If you doubt the efficacy of choosing the proper coordinate system, you should rework the last two examples in this section using spherical coordinates.

$\boxed{\vec{\nabla} \circ \vec{E}}$ The divergence of the electric field

This expression is the entire left side of the differential form of Gauss's law, and it represents the divergence of the electric field. In electrostatics, all electric field lines begin on points of positive charge and terminate on points of negative charge, so it is understandable that this expression is proportional to the electric charge density at the location under consideration.

Consider the electric field of the positive point charge; the electric field lines originate on the positive charge, and you know from Table 1.1 that the electric field is radial and decreases as $1/r^2$:

$$\vec{E} = \frac{1}{4\pi\varepsilon_0} \frac{q}{r^2} \hat{r}.$$

This is analogous to the vector field shown in Figure 1.13(c), for which the divergence is zero. Thus, the spreading out of the electric field lines is exactly compensated by the $1/r^2$ reduction in field amplitude, and the divergence of the electric field is zero at all points away from the origin.

The reason the origin (where $r = 0$) is not included in the previous analysis is that the expression for the divergence includes terms containing r in the denominator, and those terms become problematic as r approaches zero. To evaluate the divergence at the origin, use the formal definition of divergence:

$$\vec{\nabla} \circ \vec{E} \equiv \lim_{\Delta V \to 0} \frac{1}{\Delta V} \oint_S \vec{E} \circ \hat{n} \, da.$$

Considering a special Gaussian surface surrounding the point charge q, this is

$$\vec{\nabla} \circ \vec{E} \equiv \lim_{\Delta V \to 0} \left(\frac{1}{\Delta V} \frac{q}{4\pi\varepsilon_0 r^2} \oint_S da \right) = \lim_{\Delta V \to 0} \left(\frac{1}{\Delta V} \frac{q}{4\pi\varepsilon_0 r^2} (4\pi r^2) \right)$$

$$= \lim_{\Delta V \to 0} \left(\frac{1}{\Delta V} \frac{q}{\varepsilon_0} \right).$$

But $q/\Delta V$ is just the average charge density over the volume ΔV, and as ΔV shrinks to zero, this becomes equal to ρ, the charge density at the origin. Thus, at the origin the divergence is

$$\vec{\nabla} \circ \vec{E} = \frac{\rho}{\varepsilon_0},$$

in accordance with Gauss's law.

It is worth your time to make sure you understand the significance of this last point. A casual glance at the electric field lines in the vicinity of a point charge suggests that they "diverge" everywhere (in the sense of getting farther apart). But as you've seen, radial vector fields that decrease in amplitude as $1/r^2$ actually have *zero* divergence everywhere except at the source. The key factor in determining the divergence at any point is not simply the spacing of the field lines at that point, but whether the flux *out of* an infinitesimally small volume around the point is greater than, equal to, or less than the flux *into* that volume. If the outward flux exceeds the inward flux, the divergence is positive at that point. If the outward flux is less than the inward flux, the divergence is negative, and if the outward and inward fluxes are equal the divergence is zero at that point.

In the case of a point charge at the origin, the flux through an infinitesimally small surface is nonzero only if that surface contains the point charge. Everywhere else, the flux into and out of that tiny surface must be the same (since it contains no charge), and the divergence of the electric field must be zero.

$\boxed{\vec{\nabla} \circ \vec{E} = \rho/\varepsilon_0}$ **Applying Gauss's law (differential form)**

The problems you're most likely to encounter that can be solved using the differential form of Gauss's law involve calculating the divergence of the electric field and using the result to determine the charge density at a specified location.

The following examples should help you understand how to solve problems of this type.

Example 1.6: Given an expression for the vector electric field, find the divergence of the field at a specified location.

Problem: If the vector field of Figure 1.13(a) were changed to

$$\vec{A} = \sin\left(\frac{\pi}{2}y\right)\hat{i} - \sin\left(\frac{\pi}{2}x\right)\hat{j},$$

in the region $-0.5 < x < +0.5$ and $-0.5 < y < +0.5$, how would the field lines be different from those of Figure 1.13(a), and what is the divergence in this case?

Solution: When confronted with a problem like this, you may be tempted to dive in and immediately begin taking derivatives to determine the divergence of the field. A better approach is to think about the field for a moment and to attempt to *visualize* the field lines – a task that may be difficult in some cases. Fortunately, there exist a variety of computational tools such as *MATLAB*® and its freeware cousin *Octave* that are immensely helpful in revealing the details of a vector field. Using the "quiver" command in *MATLAB*® shows that the field looks as shown in Figure 1.14.

If you're surprised by the direction of the field, consider that the x-component of the field depends on y (so the field points to the right above the x-axis and to the left below the x-axis), while the y-component of the field depends on the negative of x (so the field points up on the left of the y-axis and down on the right of the y-axis). Combining these features leads to the field depicted in Figure 1.14.

Examining the field closely reveals that the flow lines neither converge nor diverge, but simply circulate back on themselves. Calculating the divergence confirms this

$$\vec{\nabla} \circ \vec{A} = \frac{\partial}{\partial x}\left[\sin\left(\frac{\pi}{2}y\right)\right] - \frac{\partial}{\partial y}\left[\sin\left(\frac{\pi}{2}x\right)\right] = 0.$$

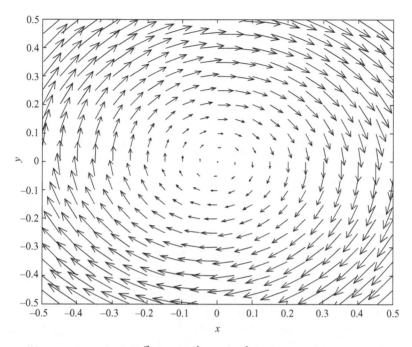

Figure 1.14 Vector field $\vec{A} = \sin(\frac{\pi}{2}y)\hat{i} - \sin(\frac{\pi}{2}x)\hat{j}$.

Electric fields that circulate back on themselves are produced not by electric charge, but rather by changing magnetic fields. Such "solenoidal" fields are discussed in Chapter 3.

Example 1.7: Given the vector electric field in a specified region, find the density of electric charge at a location within that region.

Problem: Find the charge density at $x = 2$ m and $x = 5$ m if the electric field in the region is given by

$$\vec{E} = ax^2\hat{i}\,\frac{V}{m} \quad \text{for } x = 0 \text{ to } 3 \text{ m,}$$

and

$$\vec{E} = b\hat{i}\,\frac{V}{m} \quad \text{for } x > 3 \text{ m.}$$

Solution: By Gauss's law, in the region $x = 0$ to 3 m,

$$\vec{\nabla} \circ \vec{E} = \frac{\rho}{\varepsilon_0} = \left(\hat{i}\,\frac{\partial}{\partial x} + \hat{j}\,\frac{\partial}{\partial y} + \hat{k}\,\frac{\partial}{\partial k}\right) \circ (ax^2\hat{i}),$$

$$\frac{\rho}{\varepsilon_0} = \frac{\partial(ax^2)}{\partial x} = 2xa,$$

and

$$\rho = 2xa\varepsilon_0.$$

Thus at $x = 2\,\text{m}$, $\rho = 4a\varepsilon_0$.
In the region $x > 3$ m,

$$\vec{\nabla} \circ \vec{E} = \frac{\rho}{\varepsilon_0} = \left(\hat{i}\frac{\partial}{\partial x} + \hat{j}\frac{\partial}{\partial y} + \hat{k}\frac{\partial}{\partial k} \right) \circ (b\hat{i}) = 0,$$

so $\rho = 0$ at $x = 5\,\text{m}$.

Problems
The following problems will test your understanding of Gauss's law for electric fields. Full solutions are available on the book's website.

1.1 Find the electric flux through the surface of a sphere containing 15 protons and 10 electrons. Does the size of the sphere matter?

1.2 A cube of side L contains a flat plate with variable surface charge density of $\sigma = -3xy$. If the plate extends from $x = 0$ to $x = L$ and from $y = 0$ to $y = L$, what is the total electric flux through the walls of the cube?

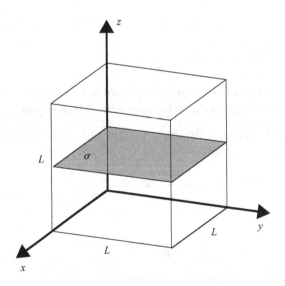

1.3 Find the total electric flux through a closed cylinder containing a line charge along its axis with linear charge density $\lambda = \lambda_0(1 - x/h)$ C/m if the cylinder and the line charge extend from $x = 0$ to $x = h$.

1.4 What is the flux through any closed surface surrounding a charged sphere of radius a_0 with volume charge density of $\rho = \rho_0(r/a_0)$, where r is the distance from the center of the sphere?

1.5 A circular disk with surface charge density $2 \times 10^{-10}\,C/m^2$ is surrounded by a sphere with radius of 1 m. If the flux through the sphere is 5.2×10^{-2} Vm, what is the diameter of the disk?

1.6 A $10\,cm \times 10\,cm$ flat plate is located 5 cm from a point charge of $10^{-8}\,C$. What is the electric flux through the plate due to the point charge?

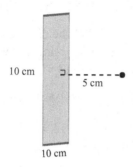

10 cm

5 cm

10 cm

1.7 Find the electric flux through a half-cylinder of height h owing to an infinitely long line charge with charge density λ running along the axis of the cylinder.

1.8 A proton rests at the center of the rim of a hemispherical bowl of radius R. What is the electric flux through the surface of the bowl?

1.9 Use a special Gaussian surface around an infinite line charge to find the electric field of the line charge as a function of distance.

1.10 Use a special Gaussian surface to prove that the magnitude of the electric field of an infinite flat plane with surface charge density σ is $|\vec{E}| = \sigma/2\varepsilon_0$.

1.11 Find the divergence of the field given by $\vec{A} = (1/r)\hat{r}$ in spherical coordinates.

1.12 Find the divergence of the field given by $\vec{A} = r\hat{r}$ in spherical coordinates.

1.13 Given the vector field

$$\vec{A} = \cos\left(\pi y - \frac{\pi}{2}\right)\hat{i} + \sin(\pi x)\hat{j},$$

sketch the field lines and find the divergence of the field.

1.14 Find the charge density in a region for which the electric field in cylindrical coordinates is given by

$$\vec{E} = \frac{az}{r}\hat{r} + br\hat{\phi} + cr^2z^2\hat{z}$$

1.15 Find the charge density in a region for which the electric field in spherical coordinates is given by

$$\vec{E} = ar^2\hat{r} + \frac{b\cos(\theta)}{r}\hat{\theta} + c\hat{\phi}.$$

2

Gauss's law for magnetic fields

Gauss's law for magnetic fields is similar in form but different in content from Gauss's law for electric fields. For both electric and magnetic fields, the integral form of Gauss's law involves the flux of the field over a closed surface, and the differential form specifies the divergence of the field at a point.

The key difference in the electric field and magnetic field versions of Gauss's law arises because opposite electric charges (called "positive" and "negative") may be isolated from one another, while opposite magnetic poles (called "north" and "south") always occur in pairs. As you might expect, the apparent lack of isolated magnetic poles in nature has a profound impact on the behavior of magnetic flux and on the divergence of the magnetic field.

2.1 The integral form of Gauss's law

Notation differs among textbooks, but the integral form of Gauss's law is generally written as follows:

$$\oint_S \vec{B} \circ \hat{n}\, da = 0 \qquad \text{Gauss's law for magnetic fields (integral form).}$$

As described in the previous chapter, the left side of this equation is a mathematical description of the flux of a vector field through a closed surface. In this case, Gauss's law refers to magnetic flux – the number of magnetic field lines – passing through a closed surface S. The right side is identically zero.

In this chapter, you will see why this law is different from the electric field case, and you will find some examples of how to use the magnetic version to solve problems – but first you should make sure you understand the main idea of Gauss's law for magnetic fields:

> The total magnetic flux passing through any closed surface is zero.

In other words, if you have a real or imaginary closed surface of any size or shape, the total magnetic flux through that surface must be zero. Note that this does not mean that zero magnetic field lines penetrate the surface – it means that for every magnetic field line that enters the volume enclosed by the surface, there must be a magnetic field line leaving that volume. Thus the inward (negative) magnetic flux must be exactly balanced by the outward (positive) magnetic flux.

Since many of the symbols in Gauss's law for magnetic fields are the same as those covered in the previous chapter, in this chapter you'll find only those symbols peculiar to this law. Here's an expanded view:

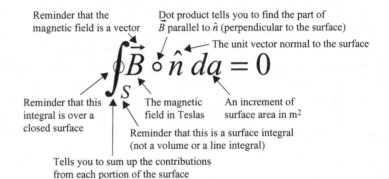

$$\oint_S \vec{B} \circ \hat{n} \, da = 0$$

Reminder that the magnetic field is a vector

Dot product tells you to find the part of $\vec{B}$ parallel to $\hat{n}$ (perpendicular to the surface)

The unit vector normal to the surface

Reminder that this integral is over a closed surface

The magnetic field in Teslas

An increment of surface area in m²

Reminder that this is a surface integral (not a volume or a line integral)

Tells you to sum up the contributions from each portion of the surface

Gauss's law for magnetic fields arises directly from the lack of isolated magnetic poles ("magnetic monopoles") in nature. Were such individual poles to exist, they would serve as the sources and sinks of magnetic field lines, just as electric charge does for electric field lines. In that case, enclosing a single magnetic pole within a closed surface would produce nonzero flux through the surface (exactly as you can produce nonzero electric flux by enclosing an electric charge). To date, all efforts to detect magnetic monopoles have failed, and every magnetic north pole is accompanied by a magnetic south pole. Thus the right side of Gauss's law for magnetic fields is identically zero.

Knowing that the total magnetic flux through a closed surface must be zero may allow you to solve problems involving complex surfaces, particularly if the flux through one portion of the surface can be found by integration.

$\vec{B}$ The magnetic field

Just as the electric field may be defined by considering the electric force on a small test charge, the magnetic field may be defined using the magnetic force experienced by a moving charged particle. As you may recall, charged particles experience magnetic force only if they are in motion with respect to the magnetic field, as shown by the Lorentz equation for magnetic force:

$$\vec{F}_B = q\,\vec{v}\times\vec{B} \qquad (2.1)$$

where $\vec{F}_B$ is the magnetic force, q is the particle's charge, $\vec{v}$ is the particle's velocity (with respect to $\vec{B}$), and $\vec{B}$ is the magnetic field.

Using the definition of the vector cross-product which says that $|\vec{a}\times\vec{b}| = |\vec{a}||\vec{b}|\sin(\theta)$, where θ is the angle between $\vec{a}$ and $\vec{b}$, the magnitude of the magnetic field may be written as

$$|\vec{B}| = \frac{|\vec{F}_B|}{q|\vec{v}|\sin(\theta)} \qquad (2.2)$$

where θ is the angle between the velocity vector $\vec{v}$ and the magnetic field $\vec{B}$. The terminology for magnetic quantities is not as standardized as that of electric quantities, so you are likely to find texts that refer to $\vec{B}$ as the "magnetic induction" or the "magnetic flux density." Whatever it is called, $\vec{B}$ has units equivalent to N/(C m/s), which include Vs/m^2, N/(Am), kg/(Cs), or most simply, Tesla (T).

Comparing Equation 2.2 to the relevant equation for the electric field, Equation (1.1), several important distinctions between magnetic and electric fields become clear:

- Like the electric field, the magnetic field is directly proportional to the magnetic force. But unlike $\vec{E}$, which is parallel or antiparallel to the electric force, the direction of $\vec{B}$ is perpendicular to the magnetic force.
- Like $\vec{E}$, the magnetic field may be defined through the force experienced by a small test charge, but unlike $\vec{E}$, the speed and direction of the test charge must be taken into consideration when relating magnetic forces and fields.
- Because the magnetic force is perpendicular to the velocity at every instant, the component of the force in the direction of the displacement is zero, and the work done by the magnetic field is therefore always zero.
- Whereas electrostatic fields are produced by electric charges, magneto-static fields are produced by electric currents.

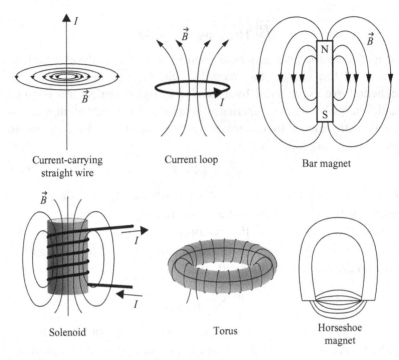

Figure 2.1 Examples of magnetic fields.

Magnetic fields may be represented using field lines whose density in a plane perpendicular to the line direction is proportional to the strength of the field. Examples of several magnetic fields relevant to the application of Gauss's law are shown in Figure 2.1.

Here are a few rules of thumb that will help you visualize and sketch the magnetic fields produced by currents:

- Magnetic field lines do not originate and terminate on charges; they form closed loops.
- The magnetic field lines that appear to originate on the north pole and terminate on the south pole of a magnet are actually continuous loops (within the magnet, the field lines run between the poles).
- The net magnetic field at any point is the vector sum of all magnetic fields present at that point.
- Magnetic field lines can never cross, since that would indicate that the field points in two different directions at the same location – if the fields from two or more sources overlap at the same location, they add (as vectors) to produce a single, total field at that point.

Table 2.1. *Magnetic field equations for simple objects*

Infinite straight wire carrying current I (at distance r)	$\vec{B} = \dfrac{\mu_0 I}{2\pi r} \hat{\varphi}$
Segment of straight wire carrying current I (at distance r)	$d\vec{B} = \dfrac{\mu_0}{4\pi} \dfrac{I d\vec{l} \times \hat{r}}{r^2}$
Circular loop of radius R carrying current I (loop in yz plane, at distance x along x-axis)	$\vec{B} = \dfrac{\mu_0 I R^2}{2(x^2 + R^2)^{3/2}} \hat{x}$
Solenoid with N turns and length l carrying current I	$\vec{B} = \dfrac{\mu_0 N I}{l} \hat{x} \ (\text{inside})$
Torus with N turns and radius r carrying current I	$\vec{B} = \dfrac{\mu_0 N I}{2\pi r} \hat{\varphi} \ (\text{inside})$

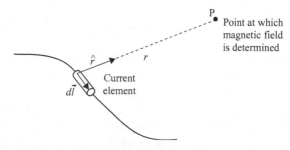

Figure 2.2 Geometry for Biot–Savart law.

All static magnetic fields are produced by moving electric charge. The contribution $d\vec{B}$ to the magnetic field at a specified point P from a small element of electric current is given by the Biot–Savart law:

$$d\vec{B} = \frac{\mu_0}{4\pi} \frac{I d\vec{l} \times \hat{r}}{r^2}$$

In this equation, μ_0 is the permeability of free space, I is the current through the small element, $d\vec{l}$ is a vector with the length of the current element and pointing in the direction of the current, $\hat{r}$ is a unit vector pointing from the current element to the point P at which the field is being calculated, and r is the distance between the current element and P, as shown in Figure 2.2.

Equations for the magnetic field in the vicinity of some simple objects may be found in Table 2.1.

$$\boxed{\oint_S \vec{B} \circ \hat{n}\, da}$$ **The magnetic flux through a closed surface**

Like the electric flux Φ_E, the magnetic flux Φ_B through a surface may be thought of as the "amount" of magnetic field "flowing" through the surface. How this quantity is calculated depends on the situation:

$$\Phi_B = |\vec{B}| \times (\text{surface area}) \quad \vec{B} \text{ uniform and perpendicular to } S, \quad (2.3)$$

$$\Phi_B = \vec{B} \circ \hat{n} \times (\text{surface area}) \quad \vec{B} \text{ uniform and at an angle to } S, \quad (2.4)$$

$$\Phi_B = \int_S \vec{B} \circ \hat{n}\, da \quad \vec{B} \text{ nonuniform and at variable angle to } S. \quad (2.5)$$

Magnetic flux, like electric flux, is a scalar quantity, and in the magnetic case, the units of flux have been given the special name "webers" (abbreviated Wb and which, by any of the relations shown above, must be equivalent to $T\,m^2$).

As in the case of electric flux, the magnetic flux through a surface may be considered to be the number of magnetic field lines penetrating that surface. When you think about the number of magnetic field lines through a surface, don't forget that magnetic fields, like electric fields, are actually continuous in space, and that "number of field lines" only has meaning once you've established a relationship between the number of lines you draw and the strength of the field.

When considering magnetic flux through a closed surface, it is especially important to remember the caveat that surface penetration is a two-way street, and that outward flux and inward flux have opposite signs. Thus equal amounts of outward (positive) flux and inward (negative) flux will cancel, producing zero net flux.

The reason that the sign of outward and inward flux is so important in the magnetic case may be understood by considering a small closed surface placed in any of the fields shown in Figure 2.1. No matter what shape of surface you choose, and no matter where in the magnetic field you place that surface, you'll find that the number of field lines entering the volume enclosed by the surface is exactly equal to the number of field lines leaving that volume. If this holds true for all magnetic fields, it can only mean that the net magnetic flux through any closed surface must always be zero.

Of course, it does hold true, because the only way to have field lines enter a volume without leaving it is to have them terminate within the

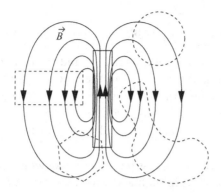

Figure 2.3 Magnetic flux lines penetrating closed surfaces.

volume, and the only way to have field lines leave a volume without entering it is to have them originate within the volume. But unlike electric field lines, magnetic field lines do not originate and terminate on charges – instead, they circulate back on themselves, forming continuous loops. If one portion of a loop passes through a closed surface, another portion of that same loop must pass through the surface in the opposite direction. Thus the outward and inward magnetic flux must be equal and opposite through any closed surface.

Consider the closer view of the field produced by a bar magnet shown in Figure 2.3. Irrespective of the shape and location of the closed surfaces placed in the field, all field lines entering the enclosed volume are offset by an equal number of field lines leaving that volume.

The physical reasoning behind Gauss's law should now be clear: the net magnetic flux passing through any closed surface must be zero because magnetic field lines always form complete loops. The next section shows you how to use this principle to solve problems involving closed surfaces and the magnetic field.

$$\boxed{\oint_S \vec{B} \circ \hat{n}\, da = 0}\ \text{Applying Gauss's law (integral form)}$$

In situations involving complex surfaces and fields, finding the flux by integrating the normal component of the magnetic field over a specified surface can be quite difficult. In such cases, knowing that the total magnetic flux through a closed surface must be zero may allow you to simplify the problem, as demonstrated by the following examples.

Example 2.1: Given an expression for the magnetic field and a surface geometry, find the flux through a specified portion of that surface.

Problem: A closed cylinder of height h and radius R is placed in a magnetic field given by $\vec{B} = B_0(\hat{j} - \hat{k})$. If the axis of the cylinder is aligned along the z-axis, find the flux through (a) the top and bottom surfaces of the cylinder and (b) the curved surface of the cylinder.

Solution: Gauss's law tells you that the magnetic flux through the entire surface must be zero, so if you're able to figure out the flux through some portions of the surface, you can deduce the flux through the other portions. In this case, the flux through the top and bottom of the cylinder are relatively easy to find; whatever additional amount it takes to make the total flux equal to zero must come from the curved sides of the cylinder. Thus

$$\Phi_{B,\text{Top}} + \Phi_{B,\text{Bottom}} + \Phi_{B,\text{Sides}} = 0.$$

The magnetic flux through any surface is

$$\Phi_B = \int_S \vec{B} \circ \hat{n}\, da.$$

For the top surface, $\hat{n} = \hat{k}$, so

$$\vec{B} \circ \hat{n} = (B_0\hat{j} - B_0\hat{k}) \circ \hat{k} = -B_0.$$

Thus

$$\Phi_{B,\text{Top}} = \int_S \vec{B} \circ \hat{n}\, da = -B_0 \int_S da = -B_0(\pi R^2).$$

A similar analysis for the bottom surface (for which $\hat{n} = -\hat{k}$) gives

$$\Phi_{B,\text{Bottom}} = \int_S \vec{B} \circ \hat{n}\, da = +B_0 \int_S da = +B_0(\pi R^2).$$

Since $\Phi_{B,\text{Top}} = -\Phi_{B,\text{Bottom}}$, you can conclude that $\Phi_{B,\text{Sides}} = 0$.

Example 2.2: Given the current in a long wire, find the magnetic flux through nearby surfaces

Problem: Find the magnetic flux through the curved surface of a half-cylinder near a long, straight wire carrying current I.

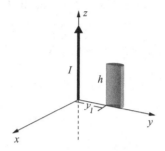

Solution: At distance r from a current-carrying wire, the magnetic field is given by

$$\vec{B} = \frac{\mu_0 I}{2\pi r} \hat{\varphi},$$

which means that the magnetic field lines make circles around the wire, entering the half-cylinder through the flat surface and leaving through the curved surface. Gauss's law tells you that the total magnetic flux through all faces of the half-cylinder must be zero, so the amount of (negative) flux through the flat surface must equal the amount of (positive) flux leaving the curved surface. To find the flux through the flat surface, use the expression for flux

$$\Phi_B = \int_S \vec{B} \circ \hat{n} \, da.$$

In this case, $\hat{n} = -\hat{\varphi}$, so

$$\vec{B} \circ \hat{n} = \left(\frac{\mu_0 I}{2\pi r} \hat{\varphi}\right) \circ (-\hat{\varphi}) = -\frac{\mu_0 I}{2\pi r}.$$

To integrate over the flat face of the half-cylinder, notice that the face lies in the yz plane, and an element of surface area is therefore $da = dy \, dz$. Notice also that on the flat face the distance increment $dr = dy$, so $da = dr \, dz$ and the flux integral is

$$\Phi_{B,\text{Flat}} = \int_S \vec{B} \circ \hat{n} \, da = -\int_S \frac{\mu_0 I}{2\pi r} \, dr \, dz = -\frac{\mu_0 I}{2\pi} \int_{z=0}^{h} \int_{r=y_1}^{y_1+2R} \frac{dr}{r} \, dz .$$

Thus

$$\Phi_{B,\text{Flat}} = -\frac{\mu_0 I}{2\pi} \ln\left(\frac{y_1 + 2R}{y_1}\right)(h) = -\frac{\mu_0 I h}{2\pi} \ln\left(1 + \frac{2R}{y_1}\right).$$

Since the total magnetic flux through this closed surface must be zero, this means that the flux through the curved side of the half-cylinder is

$$\Phi_{B,\text{Curved side}} = \frac{\mu_0 I h}{2\pi} \ln\left(1 + \frac{2R}{y_1}\right).$$

2.2 The differential form of Gauss's law

The continuous nature of magnetic field lines makes the differential form of Gauss's law for magnetic fields quite simple. The differential form is written as

$$\vec{\nabla} \circ \vec{B} = 0 \quad \text{Gauss's law for magnetic fields (differential form).}$$

The left side of this equation is a mathematical description of the divergence of the magnetic field – the tendency of the magnetic field to "flow" more strongly away from a point than toward it – while the right side is simply zero.

The divergence of the magnetic field is discussed in detail in the following section. For now, make sure you grasp the main idea of Gauss's law in differential form:

> The divergence of the magnetic field at any point is zero.

One way to understand why this is true is by analogy with the electric field, for which the divergence at any location is proportional to the electric charge density at that location. Since it is not possible to isolate magnetic poles, you can't have a north pole without a south pole, and the "magnetic charge density" must be zero everywhere. This means that the divergence of the magnetic field must also be zero.

To help you understand the meaning of each symbol in Gauss's law for magnetic fields, here is an expanded view:

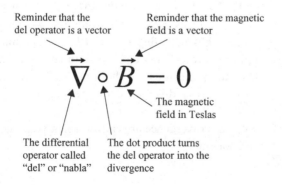

Reminder that the del operator is a vector

Reminder that the magnetic field is a vector

$$\vec{\nabla} \circ \vec{B} = 0$$

The magnetic field in Teslas

The differential operator called "del" or "nabla"

The dot product turns the del operator into the divergence

$\boxed{\vec{\nabla} \circ \vec{B}}$ The divergence of the magnetic field

This expression is the entire left side of the differential form of Gauss's law, and it represents the divergence of the magnetic field. Since divergence is by definition the tendency of a field to "flow" away from a point more strongly than toward that point, and since no point sources or sinks of the magnetic field have ever been found, the amount of "incoming" field is exactly the same as the amount of "outgoing" field at every point. So it should not surprise you to find that the divergence of $\vec{B}$ is always zero.

To verify this for the case of the magnetic field around a long, current-carrying wire, take the divergence of the expression for the wire's magnetic field as given in Table 2.1:

$$\text{div}(\vec{B}) = \vec{\nabla} \circ \vec{B} = \vec{\nabla} \circ \left(\frac{\mu_0 I}{2\pi r} \hat{\varphi} \right). \tag{2.6}$$

This is most easily determined using cylindrical coordinates:

$$\vec{\nabla} \circ \vec{B} = \frac{1}{r} \frac{\partial}{\partial r} (rB_r) + \frac{1}{r} \frac{\partial B_\phi}{\partial \phi} + \frac{\partial B_z}{\partial z}. \tag{2.7}$$

which, since $\vec{B}$ has only a φ-component, is

$$\vec{\nabla} \circ \vec{B} = \frac{1}{r} \frac{\partial (\mu_0 I / 2\pi r)}{\partial \varphi} = 0. \tag{2.8}$$

You can understand this result using the following reasoning: since the magnetic field makes circular loops around the wire, it has no z-component and no radial component (although it does have a radial dependence, since the field weakens with distance from the wire). Thus $\vec{B}_z$ and $\vec{B}_r$ both equal zero in this case, leaving only the φ-component. And since the φ-component has no φ-dependence (that is, the magnetic field has constant amplitude around any circular path centered on the wire), the flux away from any point must be the same as the flux toward that point. This means that the divergence of the magnetic field is zero everywhere.

Vector fields with zero divergence are called "solenoidal" fields, and all magnetic fields are solenoidal.

$\boxed{\vec{\nabla} \circ \vec{B} = 0}$ **Applying Gauss's law (differential form)**

Knowing that the divergence of the magnetic field must be zero allows you to attack problems involving the spatial change in the components of a magnetic field and to determine whether a specified vector field could be a magnetic field. This section has examples of such problems.

Example 2.3: Given incomplete information about the components of a magnetic field, use Gauss's law to establish relationships between those components

Problem: A magnetic field is given by the expression

$$\vec{B} = axz\hat{i} + byz\hat{j} + c\hat{k}$$

What is the relationship between a and b?

Solution: You know from Gauss's law for magnetic fields that the divergence of the magnetic field must be zero. Thus

$$\vec{\nabla} \circ \vec{B} = \frac{\partial B_x}{\partial x} + \frac{\partial B_y}{\partial y} + \frac{\partial B_z}{\partial z} = 0.$$

Thus

$$\frac{\partial(axz)}{\partial x} + \frac{\partial(byz)}{\partial y} + \frac{\partial c}{\partial z} = 0$$

and

$$az + bz + 0 = 0,$$

which means that $a = -b$.

Example 2.4: Given an expression for a vector field, determine whether that field could be a magnetic field.

Problem: A vector field is given by the expression

$$\vec{A}(x, y) = a\cos(bx)\hat{i} + aby\sin(bx)\hat{j}.$$

Could this field be a magnetic field?

Solution: Gauss's law tells you that the divergence of all magnetic fields must be zero, and checking the divergence of this vector field gives

$$\vec{\nabla} \circ \vec{A} = \frac{\partial}{\partial x}[a\cos(bx)] + \frac{\partial}{\partial y}[aby\sin(bx)]$$

$$= -ab\sin(bx) + ab\sin(bx) = 0$$

which indicates that $\vec{A}$ could represent a magnetic field.

Problems

The following problems will check your understanding of Gauss's law for magnetic fields. Full solutions are available on the book's website.

2.1 Find the magnetic flux produced by the magnetic field $\vec{B} = 5\hat{i} - 3\hat{j} + 4\hat{k}$nT through the top, bottom, and side surfaces of the flared cylinder shown in the figure.

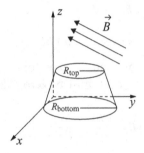

2.2 What is the change in magnetic flux through a 10 cm by 10 cm square lying 20 cm from a long wire carrying a current that increases from 5 to 15 mA? Assume that the wire is in the plane of the square and parallel to the closest side of the square.

2.3 Find the magnetic flux through all five surfaces of the wedge shown in the figure if the magnetic field in the region is given by

$$\vec{B} = 0.002\hat{i} + 0.003\hat{j}\,\text{T},$$

and show that the total flux through the wedge is zero.

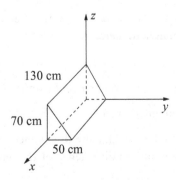

2.4 Find the flux of the Earth's magnetic field through each face of a cube with 1-m sides, and show that the total flux through the cube is zero. Assume that at the location of the cube the Earth's magnetic field has amplitude of 4×10^{-5} T and points upward at an angle of $30°$ with respect to the horizontal. You may orient the cube in any way you choose.

2.5 A cylinder of radius r_0 and height h is placed inside an ideal solenoid with the cylinder's axis parallel to the axis of the solenoid. Find the flux through the top, bottom, and curved surfaces of the cylinder and show that the total flux through the cylinder is zero.

2.6 Determine whether the vector fields given by the following expressions in cylindrical coordinates could be magnetic fields:

(a)
$$\vec{A}(r, \varphi, z) = \frac{a}{r} \cos^2(\varphi) \hat{r},$$

(b)
$$\vec{A}(r, \varphi, z) = \frac{a}{r^2} \cos^2(\varphi) \hat{r}.$$

3

Faraday's law

In a series of epoch-making experiments in 1831, Michael Faraday demonstrated that an electric current may be induced in a circuit by changing the magnetic flux enclosed by the circuit. That discovery is made even more useful when extended to the general statement that a changing magnetic field produces an electric field. Such "induced" electric fields are very different from the fields produced by electric charge, and Faraday's law of induction is the key to understanding their behavior.

3.1 The integral form of Faraday's law

In many texts, the integral form of Faraday's law is written as

$$\oint_C \vec{E} \circ d\vec{l} = -\frac{d}{dt} \int_S \vec{B} \circ \hat{n}\, da \qquad \text{Faraday's law (integral form)}.$$

Some authors feel that this form is misleading because it confounds two distinct phenomena: magnetic induction (involving a changing magnetic field) and motional electromotive force (emf) (involving movement of a charged particle through a magnetic field). In both cases, an emf is produced, but only magnetic induction leads to a circulating electric field in the rest frame of the laboratory. This means that this common version of Faraday's law is rigorously correct only with the caveat that $\vec{E}$ represents the electric field in the rest frame of each segment $d\vec{l}$ of the path of integration.

A version of Faraday's law that separates the two effects and makes clear the connection between electric field circulation and a changing magnetic field is

$$\text{emf} = -\frac{d}{dt} \int_S \vec{B} \circ \hat{n} \, da \qquad \text{Flux rule,}$$

$$\oint_C \vec{E} \circ d\vec{l} = -\int_S \frac{\partial \vec{B}}{\partial t} \circ \hat{n} \, da \qquad \text{Faraday's law (alternate form).}$$

Note that in this version of Faraday's law the time derivative operates only on the magnetic field rather than on the magnetic flux, and both $\vec{E}$ and $\vec{B}$ are measured in the laboratory reference frame.

Don't worry if you're uncertain of exactly what emf is or how it is related to the electric field; that's all explained in this chapter. There are also examples of how to use the flux rule and Faraday's law to solve problems involving induction – but first you should make sure you understand the main idea of Faraday's law:

Changing magnetic flux through a surface induces an emf in any boundary path of that surface, and a changing magnetic field induces a circulating electric field.

In other words, if the magnetic flux through a surface changes, an electric field is induced along the boundary of that surface. If a conducting material is present along that boundary, the induced electric field provides an emf that drives a current through the material. Thus quickly poking a bar magnet through a loop of wire generates an electric field within that wire, but holding the magnet in a fixed position with respect to the loop induces no electric field.

And what does the negative sign in Faraday's law tell you? Simply that the induced emf *opposes* the change in flux – that is, it tends to maintain the existing flux. This is called Lenz's law and is discussed later in this chapter.

Here's an expanded view of the standard form of Faraday's law:

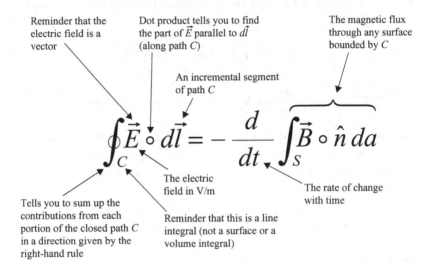

Note that $\vec{E}$ in this expression is the induced electric field at each segment $d\vec{l}$ of the path C measured in the reference frame in which that segment is stationary.

And here is an expanded view of the alternative form of Faraday's law:

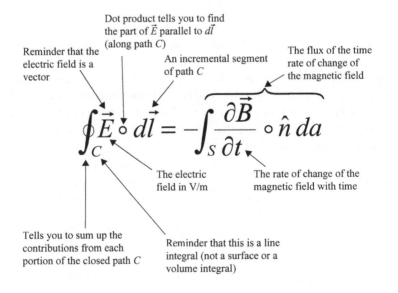

In this case, $\vec{E}$ represents the electric field in the laboratory frame of reference (the same frame in which $\vec{B}$ is measured).

Faraday's law and the flux rule can be used to solve a variety of problems involving changing magnetic flux and induced electric fields, in particular problems of two types:

(1) Given information about the changing magnetic flux, find the induced emf.
(2) Given the induced emf on a specified path, determine the rate of change of the magnetic field magnitude or direction or the area bounded by the path.

In situations of high symmetry, in addition to finding the induced emf, it is also possible to find the induced electric field when the rate of change of the magnetic field is known.

$\vec{E}$ The induced electric field

The electric field in Faraday's law is similar to the electrostatic field in its *effect* on electric charges, but quite different in its *structure*. Both types of electric field accelerate electric charges, both have units of N/C or V/m, and both can be represented by field lines. But charge-based electric fields have field lines that originate on positive charge and terminate on negative charge (and thus have non-zero divergence at those points), while induced electric fields produced by changing magnetic fields have field lines that loop back on themselves, with no points of origination or termination (and thus have zero divergence).

It is important to understand that the electric field that appears in the common form of Faraday's law (the one with the full derivative of the magnetic flux on the right side) is the electric field measured in the reference frame of each segment $d\vec{l}$ of the path over which the circulation is calculated. The reason for making this distinction is that it is only in this frame that the electric field lines actually circulate back on themselves.

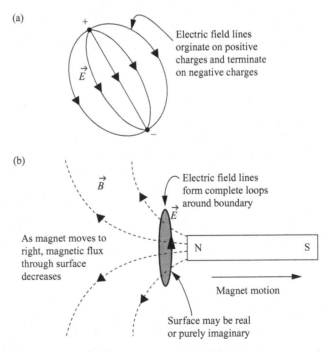

(a)

Electric field lines orginate on positive charges and terminate on negative charges

$\vec{E}$

(b)

$\vec{B}$

Electric field lines form complete loops around boundary

$\vec{E}$

As magnet moves to right, magnetic flux through surface decreases

N   S

Magnet motion

Surface may be real or purely imaginary

Figure 3.1 Charge-based and induced electric fields. As always, you should remember that these fields exist in three dimensions, and you can see full 3-D visualizations on the book's website.

Examples of a charge-based and an induced electric field are shown in Figure 3.1.

Note that the induced electric field in Figure 3.1(b) is directed so as to drive an electric current that produces magnetic flux that opposes the change in flux due to the changing magnetic field. In this case, the motion of the magnet to the right means that the leftward magnetic flux is decreasing, so the induced current produces additional leftward magnetic flux.

Here are a few rules of thumb that will help you visualize and sketch the electric fields produced by changing magnetic fields:

- Induced electric field lines produced by changing magnetic fields must form complete loops.
- The net electric field at any point is the vector sum of all electric fields present at that point.
- Electric field lines can never cross, since that would indicate that the field points in two different directions at the same location.

In summary, the $\vec{E}$ in Faraday's law represents the induced electric field at each point along path C, a boundary of the surface through which the magnetic flux is changing over time. The path may be through empty space or through a physical material – the induced electric field exists in either case.

$$\oint_C (\;)dl \quad \text{The line integral}$$

To understand Faraday's law, it is essential that you comprehend the meaning of the line integral. This type of integral is common in physics and engineering, and you have probably come across it before, perhaps when confronted with a problem such as this: find the total mass of a wire for which the density varies along its length. This problem serves as a good review of line integrals.

Consider the variable-density wire shown in Figure 3.2(a). To determine the total mass of the wire, imagine dividing the wire into a series of short segments over each of which the linear density λ (mass per unit length) is approximately constant, as shown in Figure 3.2(b). The mass of each segment is the product of the linear density of that segment times the segment length dx_i, and the mass of the entire wire is the sum of the segment masses.

For N segments, this is

$$\text{Mass} = \sum_{i=1}^{N} \lambda_i \, dx_i. \qquad (3.1)$$

Allowing the segment length to approach zero turns the summation of the segment masses into a line integral:

$$\text{Mass} = \int_0^L \lambda(x) \, dx. \qquad (3.2)$$

This is the line integral of the scalar function $\lambda(x)$. To fully comprehend the left side of Faraday's law, you'll have to understand how to extend this concept to the path integral of a vector field, which you can read about in the next section.

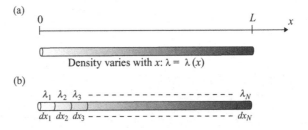

Figure 3.2 Line integral for a scalar function.

$\oint_C \vec{A} \circ d\vec{l}$ The path integral of a vector field

The line integral of a vector field around a closed path is called the "circulation" of the field. A good way to understand the meaning of this operation is to consider the work done by a force as it moves an object along a path.

As you may recall, work is done when an object is displaced under the influence of a force. If the force $(\vec{F})$ is constant and in the same direction as the displacement $(d\vec{l})$, the amount of work (W) done by the force is simply the product of the magnitudes of the force and the displacement:

$$W = |\vec{F}| \, |d\vec{l}|. \tag{3.3}$$

This situation is illustrated in Figure 3.3(a). In many cases, the displacement is not in the same direction as the force, and it then becomes necessary to determine the component of the force in the direction of the displacement, as shown in Figure 3.3(b).

In this case, the amount of work done by the force is equal to the component of the force in the direction of the displacement multiplied by the amount of displacement. This is most easily signified using the dot product notation described in Chapter 1:

$$W = \vec{F} \circ d\vec{l} = |\vec{F}||d\vec{l}| \cos(\theta), \tag{3.4}$$

where θ is the angle between the force and the displacement.

In the most general case, the force $\vec{F}$ and the angle between the force and the displacement may not be constant, which means that the projection of the force on each segment may be different (it is also possible that the magnitude of the force may change along the path). The general case is illustrated in Figure 3.4. Note that as the path meanders from the starting point to the end, the component of the force in the direction of the displacement varies considerably.

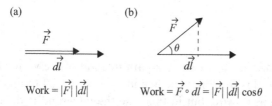

Figure 3.3 Object moving under the influence of a force.

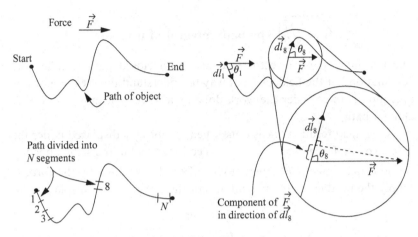

Figure 3.4 Component of force along object path.

To find the work in this case, the path may be thought of as a series of short segments over each of which the component of the force is constant. The incremental work (dW_i) done over each segment is simply the component of the force along the path at that segment times the segment length (dl_i) – and that's exactly what the dot product does. Thus,

$$dW_i = \vec{F} \circ \vec{dl_i}, \qquad (3.5)$$

and the work done along the entire path is then just the summation of the incremental work done at each segment, which is

$$W = \sum_{i=1}^{N} dW_i = \sum_{i=1}^{N} \vec{F} \circ \vec{dl_i}. \qquad (3.6)$$

As you've probably guessed, you can now allow the segment length to shrink toward zero, converting the sum to an integral over the path:

$$W = \int_{P} \vec{F} \circ \vec{dl}. \qquad (3.7)$$

Thus, the work in this case is the path integral of the vector $\vec{F}$ over path P. This integral is similar to the line integral you used to find the mass of a variable-density wire, but in this case the integrand is the dot product between two vectors rather than the scalar function λ.

Although the force in this example is uniform, the same analysis pertains to a vector field of force that varies in magnitude and direction along the path. The integral on the right side of Equation 3.7 may be defined for any vector field $\vec{A}$ and any path C. If the path is closed, this integral represents the circulation of the vector field around that path:

$$\text{Circulation} \equiv \oint_C \vec{A} \circ \vec{dl}. \qquad (3.8)$$

The circulation of the electric field is an important part of Faraday's law, as described in the next section.

$\boxed{\oint_C \vec{E} \circ \vec{dl}}$ The electric field circulation

Since the field lines of induced electric fields form closed loops, these fields are capable of driving charged particles around continuous circuits. Charge moving through a circuit is the very definition of electric current, so the induced electric field may act as a generator of electric current. It is therefore understandable that the circulation of the electric field around a circuit has come to be known as an "electromotive force":

$$\text{electromotive force (emf)} = \oint_C \vec{E} \circ \vec{dl}. \tag{3.9}$$

Of course, the path integral of an electric field is not a force (which must have SI units of newtons), but rather a force per unit charge integrated over a distance (with units of newtons per coulomb times meters, which are the same as volts). Nonetheless, the terminology is now standard, and "source of emf" is often applied to induced electric fields as well as to batteries and other sources of electrical energy.

So, exactly what is the circulation of the induced electric field around a path? It is just the work done by the electric field in moving a unit charge around that path, as you can see by substituting $\vec{F}/q$ for $\vec{E}$ in the circulation integral:

$$\oint_C \vec{E} \circ \vec{dl} = \oint_C \frac{\vec{F}}{q} \circ \vec{dl} = \frac{\oint_C \vec{F} \circ \vec{dl}}{q} = \frac{W}{q}. \tag{3.10}$$

Thus, the circulation of the induced electric field is the energy given to each coulomb of charge as it moves around the circuit.

$$\boxed{\frac{d}{dt}\int_S \vec{B} \circ \hat{n}\ da}\ \ \text{The rate of change of flux}$$

The right side of the common form of Faraday's law may look intimidating at first glance, but a careful inspection of the terms reveals that the largest portion of this expression is simply the magnetic flux (Φ_B):

$$\Phi_B = \int_S \vec{B} \circ \hat{n}\ da.$$

If you're tempted to think that this quantity must be zero according to Gauss's law for magnetic fields, look more carefully. The integral in this expression is over any surface S, whereas the integral in Gauss's law is specifically over a *closed* surface. The magnetic flux (proportional to the number of magnetic field lines) through an open surface may indeed be nonzero – it is only when the surface is closed that the number of magnetic field lines passing through the surface in one direction must equal the number passing through in the other direction.

So the right side of this form of Faraday's law involves the magnetic flux through any surface S – more specifically, the rate of change with time (d/dt) of that flux. If you're wondering how the magnetic flux through a surface might change, just look at the equation and ask yourself what might vary with time in this expression. Here are three possibilities, each of which is illustrated in Figure 3.5:

- The magnitude of $\vec{B}$ might change: the strength of the magnetic field may be increasing or decreasing, causing the number of field lines penetrating the surface to change.
- The angle between $\vec{B}$ and the surface normal might change: varying the direction of either $\vec{B}$ or the surface normal causes $\vec{B} \circ \hat{n}$ to change.
- The area of the surface might change: even if the magnitude of $\vec{B}$ and the direction of both $\vec{B}$ and $\hat{n}$ remain the same, varying the area of surface S will change the value of the flux through the surface.

Each of these changes, or a combination of them, causes the right side of Faraday's law to become nonzero. And since the left side of Faraday's law is the induced emf, you should now understand the relationship between induced emf and changing magnetic flux.

To connect the mathematical statement of Faraday's law to physical effects, consider the magnetic fields and conducting loops shown in Figure 3.5. As Faraday discovered, the mere presence of magnetic flux

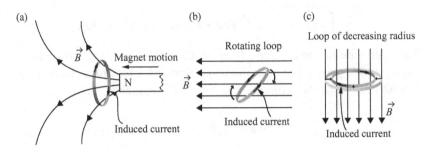

Figure 3.5 Magnetic flux and induced current.

through a circuit does not produce an electric current within that circuit. Thus, holding a stationary magnet near a stationary conducting loop induces no current (in this case, the magnetic flux is not a function of time, so its time derivative is zero and the induced emf must also be zero).

Of course, Faraday's law tells you that *changing* the magnetic flux through a surface does induce an emf in any circuit that is a boundary to that surface. So, moving a magnet toward or away from the loop, as in Figure 3.5(a), causes the magnetic flux through the surface bounded by the loop to change, resulting in an induced emf around the circuit.[4]

In Figure 3.5(b), the change in magnetic flux is produced not by moving the magnet, but by rotating the loop. This changes the angle between the magnetic field and the surface normal, which changes $\vec{B} \circ \hat{n}$. In Figure 3.5(c), the area enclosed by the loop is changing over time, which changes the flux through the surface. In each of these cases, you should note that the magnitude of the induced emf does not depend on the total amount of magnetic flux through the loop – it depends only on how fast the flux changes.

Before looking at some examples of how to use Faraday's law to solve problems, you should consider the *direction* of the induced electric field, which is provided by Lenz's law.

---

[4] For simplicity, you can imagine a planar surface stretched across the loop, but Faraday's law holds for any surface bounded by the loop.

⊟ Lenz's law

There's a great deal of physics wrapped up in the minus sign on the right side of Faraday's law, so it is fitting that it has a name: Lenz's law. The name comes from Heinrich Lenz, a German physicist who had an important insight concerning the *direction* of the current induced by changing magnetic flux.

Lenz's insight was this: currents induced by changing magnetic flux always flow in the direction so as to *oppose* the change in flux. That is, if the magnetic flux through the circuit is increasing, the induced current produces its own magnetic flux in the opposite direction to offset the increase. This situation is shown in Figure 3.6(a), in which the magnet is moving toward the loop. As the leftward flux due to the magnet increases, the induced current flows in the direction shown, which produces rightward magnetic flux that opposes the increased flux from the magnet.

The alternative situation is shown in Figure 3.6(b), in which the magnet is moving away from the loop and the leftward flux through the circuit is decreasing. In this case, the induced current flows in the opposite direction, contributing leftward flux to make up for the decreasing flux from the magnet.

It is important for you to understand that changing magnetic flux induces an electric field whether or not a conducting path exists in which a current may flow. Thus, Lenz's law tells you the direction of the circulation of the induced electric field around a specified path even if no conduction current actually flows along that path.

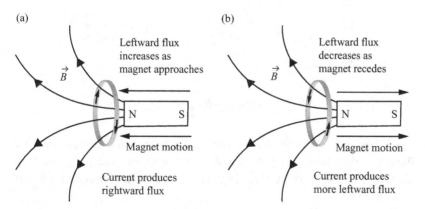

Figure 3.6 Direction of induced current.

$$\boxed{\oint_C \vec{E} \circ d\vec{l} = -\frac{d}{dt} \int_S \vec{B} \circ \hat{n}\, da}$$ **Applying Faraday's law (integral form)**

The following examples show you how to use Faraday's law to solve problems involving changing magnetic flux and induced emf.

Example 3.1: Given an expression for the magnetic field as a function of time, determine the emf induced in a loop of specified size.

Problem: For a magnetic field given by

$$\vec{B}(y,t) = B_0 \left(\frac{t}{t_0}\right) \frac{y}{y_0}\, \hat{z}.$$

Find the emf induced in a square loop of side L lying in the xy-plane with one corner at the origin. Also, find the direction of current flow in the loop.

Solution: Using Faraday's flux rule,

$$\text{emf} = -\frac{d}{dt} \int_S \vec{B} \circ \hat{n}\, da.$$

For a loop in the xy-plane, $\hat{n} = \hat{z}$ and $da = dx\, dy$, so

$$\text{emf} = -\frac{d}{dt} \int_{y=0}^{L} \int_{x=0}^{L} B_0 \left(\frac{t}{t_0}\right) \frac{y}{y_0} \hat{z} \circ \hat{z}\, dx\, dy,$$

and

$$\text{emf} = -\frac{d}{dt} \left[L \int_{y=0}^{L} B_0 \left(\frac{t}{t_0}\right) \frac{y}{y_0} dy \right] = -\frac{d}{dt} \left[B_0 \left(\frac{t}{t_0}\right) \frac{L^3}{2y_0} \right].$$

Taking the time derivative gives

$$\text{emf} = -B_0 \frac{L^3}{2t_0 y_0}.$$

Since upward magnetic flux is increasing with time, the current will flow in a direction that produces flux in the downward ($-\hat{z}$) direction. This means the current will flow in the clockwise direction as seen from above.

Example 3.2: Given an expression for the change in orientation of a conducting loop in a fixed magnetic field, find the emf induced in the loop.

Problem: A circular loop of radius r_0 rotates with angular speed ω in a fixed magnetic field as shown in the figure.

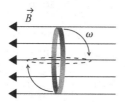

(a) Find an expression for the emf induced in the loop.
(b) If the magnitude of the magnetic field is 25 μT, the radius of the loop is 1 cm, the resistance of the loop is 25 Ω, and the rotation rate ω is 3 rad/s, what is the maximum current in the loop?

Solution: (a) By Faraday's flux rule, the emf is

$$\text{emf} = -\frac{d}{dt} \int_S \vec{B} \circ \hat{n} \, da$$

Since the magnetic field and the area of the loop are constant, this becomes

$$\text{emf} = -\int_S \frac{d}{dt}(\vec{B} \circ \hat{n}) \, da = -\int_S |\vec{B}| \frac{d}{dt} (\cos\theta) \, da.$$

Using $\theta = \omega t$, this is

$$\text{emf} = -\int_S |\vec{B}| \frac{d}{dt} (\cos\omega t) \, da = -|\vec{B}| \frac{d}{dt} (\cos\omega t) \int_S da.$$

Taking the time derivative and performing the integration gives

$$\text{emf} = |\vec{B}| \omega (\sin\omega t)(\pi r_0^2).$$

(b) By Ohm's law, the current is the emf divided by the resistance of the circuit, which is

$$I = \frac{\text{emf}}{R} = \frac{|\vec{B}| \omega (\sin\omega t)(\pi r_0^2)}{R}.$$

For maximum current, $\sin(\omega t) = 1$, so the current is

$$I = \frac{(25 \times 10^{-6})(3)[\pi(0.01^2)]}{25} = 9.4 \times 10^{-10} \, \text{A}.$$

Example 3.3: Given an expression for the change in size of a conducting loop in a fixed magnetic field, find the emf induced in the loop.

Problem: A circular loop lying perpendicular to a fixed magnetic field decreases in size over time. If the radius of the loop is given by $r(t) = r_0(1-t/t_0)$, find the emf induced in the loop.

Solution: Since the loop is perpendicular to the magnetic field, the loop normal is parallel to $\vec{B}$, and Faraday's flux rule is

$$\text{emf} = -\frac{d}{dt}\int_S \vec{B} \circ \hat{n}\, da = -|\vec{B}|\frac{d}{dt}\int_S da = -|\vec{B}|\frac{d}{dt}(\pi r^2).$$

Inserting $r(t)$ and taking the time derivative gives

$$\text{emf} = -|\vec{B}|\frac{d}{dt}\left[\pi r_0^2\left(1-\frac{t}{t_0}\right)^2\right] = -|\vec{B}|\left[\pi r_0^2(2)\left(1-\frac{t}{t_0}\right)\left(-\frac{1}{t_0}\right)\right],$$

or

$$\text{emf} = \frac{2|\vec{B}|\pi r_0^2}{t_0}\left(1-\frac{t}{t_0}\right).$$

3.2 The differential form of Faraday's law

The differential form of Faraday's law is generally written as

$$\vec{\nabla} \times \vec{E} = -\frac{\partial \vec{B}}{\partial t} \quad \text{Faraday's law.}$$

The left side of this equation is a mathematical description of the curl of the electric field – the tendency of the field lines to circulate around a point. The right side represents the rate of change of the magnetic field over time.

The curl of the electric field is discussed in detail in the following section. For now, make sure you grasp the main idea of Faraday's law in differential form:

> A circulating electric field is produced by a magnetic field that changes with time.

To help you understand the meaning of each symbol in the differential form of Faraday's law, here's an expanded view:

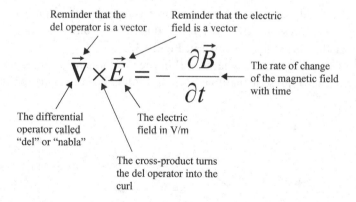

$\boxed{\vec{\nabla} \times}$ Del cross – the curl

The curl of a vector field is a measure of the field's tendency to circulate about a point – much like the divergence is a measure of the tendency of the field to flow away from a point. Once again we have Maxwell to thank for the terminology; he settled on "curl" after considering several alternatives, including "turn" and "twirl" (which he thought was somewhat racy).

Just as the divergence is found by considering the flux through an infinitesimal surface surrounding the point of interest, the curl at a specified point may be found by considering the circulation per unit area over an infinitesimal path around that point. The mathematical definition of the curl of a vector field $\vec{A}$ is

$$\text{curl}(\vec{A}) \circ \hat{n} = (\vec{\nabla} \times \vec{A}) \circ \hat{n} \equiv \lim_{\Delta S \to 0} \frac{1}{\Delta S} \oint_C \vec{A} \circ d\vec{l}, \qquad (3.11)$$

where C is a path around the point of interest and ΔS is the surface area enclosed by that path. In this definition, the direction of the curl is the normal direction of the surface for which the circulation is a maximum.

This expression is useful in defining the curl, but it doesn't offer much help in actually calculating the curl of a specified field. You'll find an alternative expression for curl later in this section, but first you should consider the vector fields shown in Figure 3.7.

To find the locations of large curl in each of these fields, imagine that the field lines represent the flow lines of a fluid. Then look for points at

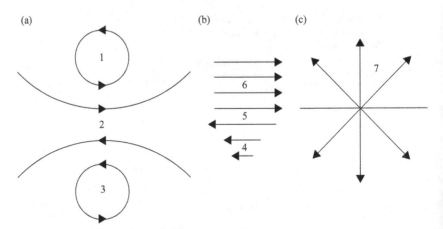

Figure 3.7 Vector fields with various values of curl.

which the flow vectors on one side of the point are significantly different (in magnitude, direction, or both) from the flow vectors on the opposite side of the point.

To aid this thought experiment, imagine holding a tiny paddlewheel at each point in the flow. If the flow would cause the paddlewheel to rotate, the center of the wheel marks a point of nonzero curl. The direction of the curl is along the axis of the paddlewheel (as a vector, curl must have both magnitude and direction). By convention, the positive-curl direction is determined by the right-hand rule: if you curl the fingers of your right hand along the circulation, your thumb points in the direction of positive curl.

Using the paddlewheel test, you can see that points 1, 2, and 3 in Figure 3.7(a) and points 4 and 5 in Figure 3.7(b) are high-curl locations. The uniform flow around point 6 in Figure 3.7(b) and the diverging flow lines around point 7 in Figure 3.7(c) would not cause a tiny paddlewheel to rotate, meaning that these are points of low or zero curl.

To make this quantitative, you can use the differential form of the curl or "del cross" ($\vec{\nabla}\times$) operator in Cartesian coordinates:

$$\vec{\nabla}\times\vec{A} = \left(\hat{i}\frac{\partial}{\partial x} + \hat{j}\frac{\partial}{\partial y} + \hat{k}\frac{\partial}{\partial z}\right) \times (\hat{i}A_x + \hat{j}A_y + \hat{k}A_z). \qquad (3.12)$$

The vector cross-product may be written as a determinant:

$$\vec{\nabla}\times\vec{A} = \begin{vmatrix} \hat{i} & \hat{j} & \hat{k} \\ \frac{\partial}{\partial x} & \frac{\partial}{\partial y} & \frac{\partial}{\partial z} \\ A_x & A_y & A_z \end{vmatrix}, \qquad (3.13)$$

which expands to

$$\vec{\nabla}\times\vec{A} = \left(\frac{\partial A_z}{\partial y} - \frac{\partial A_y}{\partial z}\right)\hat{i} + \left(\frac{\partial A_x}{\partial z} - \frac{\partial A_z}{\partial x}\right)\hat{j} + \left(\frac{\partial A_y}{\partial x} - \frac{\partial A_x}{\partial y}\right)\hat{k}. \qquad (3.14)$$

Note that each component of the curl of $\vec{A}$ indicates the tendency of the field to rotate in one of the coordinate planes. If the curl of the field at a point has a large x-component, it means that the field has significant circulation about that point in the y–z plane. The overall direction of the curl represents the axis about which the rotation is greatest, with the sense of the rotation given by the right-hand rule.

If you're wondering how the terms in this equation measure rotation, consider the vector fields shown in Figure 3.8. Look first at the field in

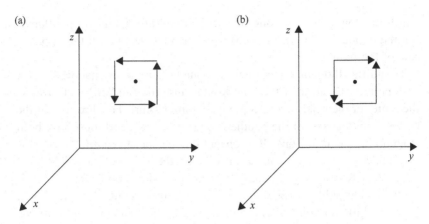

Figure 3.8 Effect of $\partial A_y/\partial z$ and $\partial A_z/\partial Y$ on value of curl.

Figure 3.8(a) and the x-component of the curl in the equation: this term involves the change in A_z with y and the change in A_y with z. Proceeding along the y-axis from the left side of the point of interest to the right, A_z is clearly increasing (it is negative on the left side of the point of interest and positive on the right side), so the term $\partial A_z/\partial y$ must be positive. Looking now at A_y, you can see that it is positive below the point of interest and negative above, so it is decreasing along the z axis. Thus, $\partial A_y/\partial z$ is negative, which means that it increases the value of the curl when it is subtracted from $\partial A_z/\partial y$. Thus the curl has a large value at the point of interest, as expected in light of the circulation of $\vec{A}$ about this point.

The situation in Figure 3.8(b) is quite different. In this case, both $\partial A_y/\partial z$ and $\partial A_z/\partial y$ are positive, and subtracting $\partial A_y/\partial z$ from $\partial A_z/\partial y$ gives a small result. The value of the x-component of the curl is therefore small in this case. Vector fields with zero curl at all points are called "irrotational."

Here are expressions for the curl in cylindrical and spherical coordinates:

$$\nabla \times \vec{A} \equiv \left(\frac{1}{r}\frac{\partial A_z}{\partial \varphi} - \frac{\partial A_\varphi}{\partial z}\right)\hat{r} + \left(\frac{\partial A_r}{\partial z} - \frac{\partial A_z}{\partial r}\right)\hat{\varphi}$$
$$+ \frac{1}{r}\left(\frac{\partial(rA_\varphi)}{\partial r} - \frac{\partial A_r}{\partial \varphi}\right)\hat{z} \quad \text{(cylindrical)},$$

(3.15)

$$\nabla \times \vec{A} \equiv \left(\frac{1}{r\sin\theta}\frac{\partial(A_\varphi \sin\theta)}{\partial \theta} - \frac{\partial A_\theta}{\partial \varphi}\right)\hat{r} + \frac{1}{r}\left(\frac{1}{\sin\theta}\frac{\partial A_r}{\partial \varphi} - \frac{\partial(rA_\varphi)}{\partial r}\right)\hat{\theta}$$
$$+ \frac{1}{r}\left(\frac{\partial(rA_\theta)}{\partial r} - \frac{\partial A_r}{\partial \theta}\right)\hat{\varphi} \quad \text{(spherical)}.$$

(3.16)

$\boxed{\vec{\nabla} \times \vec{E}}$ The curl of the electric field

Since charge-based electric fields diverge away from points of positive charge and converge toward points of negative charge, such fields cannot circulate back on themselves. You can understand that by looking at the field lines for the electric dipole shown in Figure 3.9(a). Imagine moving along a closed path that follows one of the electric field lines diverging from the positive charge, such as the dashed line shown in the figure. To close the loop and return to the positive charge, you'll have to move "upstream" against the electric field for a portion of the path. For that segment, $\vec{E} \circ \vec{dl}$ is negative, and the contribution from this part of the path subtracts from the positive value of $\vec{E} \circ \vec{dl}$ for the portion of the path in which $\vec{E}$ and $\vec{dl}$ are in the same direction. Once you've gone all the way around the loop, the integration of $\vec{E} \circ \vec{dl}$ yields exactly zero.

Thus, the field of the electric dipole, like all electrostatic fields, has no curl.

Electric fields induced by changing magnetic fields are very different, as you can see in Figure 3.9(b). Wherever a changing magnetic field exists, a circulating electric field is induced. Unlike charge-based electric fields, induced fields have no origination or termination points – they are continuous and circulate back on themselves. Integrating $\vec{E} \circ \vec{dl}$ around any boundary path for the surface through which $\vec{B}$ is changing produces a nonzero result, which means that induced electric fields have curl. The faster $\vec{B}$ changes, the larger the magnitude of the curl of the induced electric field.

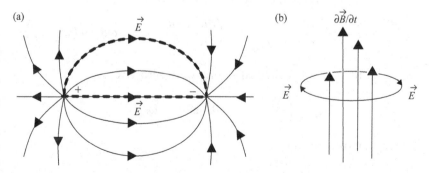

Figure 3.9 Closed paths in charge-based and induced electric fields.

$$\boxed{\vec{\nabla}\times\vec{E} = -\frac{\partial\vec{B}}{\partial t}}$$ **Applying Faraday's law (differential form)**

The differential form of Faraday's law is very useful in deriving the electromagnetic wave equation, which you can read about in Chapter 5. You may also encounter two types of problems that can be solved using this equation. In one type, you're provided with an expression for the magnetic field as a function of time and asked to find the curl of the induced electric field. In the other type, you're given an expression for the induced vector electric field and asked to determine the time rate of change of the magnetic field. Here are two examples of such problems.

Example 3.4: Given an expression for the magnetic field as a function of time, find the curl of the electric field.

Problem: The magnetic field in a certain region is given by the expression $\vec{B}(t) = B_0 \cos(kz - \omega t)\hat{j}$.

(a) Find the curl of the induced electric field at that location.
(b) If the E_z is known to be zero, find E_x.

Solution: (a) By Faraday's law, the curl of the electric field is the negative of the derivative of the vector magnetic field with respect to time. Thus,

$$\vec{\nabla}\times\vec{E} = -\frac{\partial\vec{B}}{\partial t} = -\frac{\partial[B_0\cos(kz-\omega t)]\hat{j}}{\partial t},$$

or

$$\vec{\nabla}\times\vec{E} = -\omega B_0 \sin(kz - \omega t)\hat{j}.$$

(b) Writing out the components of the curl gives

$$\left(\frac{\partial E_z}{\partial y} - \frac{\partial E_y}{\partial z}\right)\hat{i} + \left(\frac{\partial E_x}{\partial z} - \frac{\partial E_z}{\partial x}\right)\hat{j} + \left(\frac{\partial E_y}{\partial x} - \frac{\partial E_x}{\partial y}\right)\hat{k} = -\omega B_0 \sin(kz - \omega t)\hat{j}.$$

Equating the $\hat{j}$ components and setting E_z to zero gives

$$\left(\frac{\partial E_x}{\partial z}\right) = -\omega B_0 \sin(kz - \omega t).$$

Integrating over z gives

$$E_x = \int -\omega B_0 \sin(kz - \omega t)dz = \frac{\omega}{k}B_0 \cos(kz - \omega t),$$

to within a constant of integration.

Example 3.5: Given an expression for the induced electric field, find the time rate of change of the magnetic field.

Problem: Find the rate of change with time of the magnetic field at a location at which the induced electric field is given by

$$\vec{E}(x,y,z) = E_0\left[\left(\frac{z}{z_0}\right)^2\hat{i} + \left(\frac{x}{x_0}\right)^2\hat{j} + \left(\frac{y}{y_0}\right)^2\hat{k}\right].$$

Solution: Faraday's law tells you that the curl of the induced electric field is equal to the negative of the time rate of change of the magnetic field. Thus

$$\frac{\partial\vec{B}}{\partial t} = -\vec{\nabla}\times\vec{E},$$

which in this case gives

$$\frac{\partial\vec{B}}{\partial t} = -\left(\frac{\partial E_z}{\partial y} - \frac{\partial E_y}{\partial z}\right)\hat{i} - \left(\frac{\partial E_x}{\partial z} - \frac{\partial E_z}{\partial x}\right)\hat{j} - \left(\frac{\partial E_y}{\partial x} - \frac{\partial E_x}{\partial y}\right)\hat{k},$$

$$\frac{\partial\vec{B}}{\partial t} = -E_0\left[\left(\frac{2y}{y_0^2}\right)\hat{i} + \left(\frac{2z}{z_0^2}\right)\hat{j} + \left(\frac{2x}{x_0^2}\right)\hat{k}\right].$$

Problems

You can exercise your understanding of Faraday's law on the following problems. Full solutions are available on the book's website.

3.1 Find the emf induced in a square loop with sides of length a lying in the yz-plane in a region in which the magnetic field changes over time as $\vec{B}(t) = B_0 e^{-5t/t_0}\hat{i}$.

3.2 A square conducting loop with sides of length L rotates so that the angle between the normal to the plane of the loop and a fixed magnetic field $\vec{B}$ varies as $\theta(t) = \theta_0(t/t_0)$; find the emf induced in the loop.

3.3 A conducting bar descends with speed v down conducting rails in the presence of a constant, uniform magnetic field pointing into the page, as shown in the figure.

(a) Write an expression for the emf induced in the loop.

(b) Determine the direction of current flow in the loop.

3.4 A square loop of side *a* moves with speed *v* into a region in which a magnetic field of magnitude B_0 exists perpendicular to the plane of the loop, as shown in the figure. Make a plot of the emf induced in the loop as it enters, moves through, and exits the region of the magnetic field.

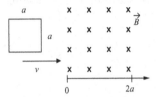

3.5 A circular loop of wire of radius 20 cm and resistance of 12 Ω surrounds a 5-turn solenoid of length 38 cm and radius 10 cm, as shown in the figure. If the current in the solenoid increases linearly from 80 to 300 mA in 2 s, what is the maximum current induced in the loop?

3.6 A 125-turn rectangular coil of wire with sides of 25 and 40 cm rotates about a horizontal axis in a vertical magnetic field of 3.5 mT. How fast must this coil rotate for the induced emf to reach 5V?

3.7 The current in a long solenoid varies as $I(t) = I_0 \sin(\omega t)$. Use Faraday's law to find the induced electric field as a function of *r* both inside and outside the solenoid, where *r* is the distance from the axis of the solenoid.

3.8 The current in a long, straight wire decreases as $I(t) = I_0 e^{-t/\tau}$. Find the induced emf in a square loop of wire of side *s* lying in the plane of the current-carrying wire at a distance *d*, as shown in the figure.

4

The Ampere–Maxwell law

For thousands of years, the only known sources of magnetic fields were certain iron ores and other materials that had been accidentally or deliberately magnetized. Then in 1820, French physicist Andre-Marie Ampere heard that in Denmark Hans Christian Oersted had deflected a compass needle by passing an electric current nearby, and within one week Ampere had begun quantifying the relationship between electric currents and magnetic fields.

"Ampere's law" relating a steady electric current to a circulating magnetic field was well known by the time James Clerk Maxwell began his work in the field in the 1850s. However, Ampere's law was known to apply only to static situations involving steady currents. It was Maxwell's addition of another source term – a changing electric flux – that extended the applicability of Ampere's law to time-dependent conditions. More importantly, it was the presence of this term in the equation now called the Ampere–Maxwell law that allowed Maxwell to discern the electromagnetic nature of light and to develop a comprehensive theory of electromagnetism.

4.1 The integral form of the Ampere–Maxwell law

The integral form of the Ampere–Maxwell law is generally written as

$$\oint_C \vec{B} \circ d\vec{l} = \mu_0 \left(I_{\text{enc}} + \varepsilon_0 \frac{d}{dt} \int_S \vec{E} \circ \hat{n} \, da \right) \quad \text{The Ampere–Maxwell law.}$$

The left side of this equation is a mathematical description of the circulation of the magnetic field around a closed path C. The right side includes two sources for the magnetic field; a steady conduction current and a changing electric flux through any surface S bounded by path C.

In this chapter, you'll find a discussion of the circulation of the magnetic field, a description of how to determine which current to include in calculating $\vec{B}$, and an explanation of why the changing electric flux is called the "displacement current." There are also examples of how to use the Ampere–Maxwell law to solve problems involving currents and magnetic fields. As always, you should begin by reviewing the main idea of the Ampere–Maxwell law:

> An electric current or a changing electric flux through a surface produces a circulating magnetic field around any path that bounds that surface.

In other words, a magnetic field is produced along a path if any current is enclosed by the path or if the electric flux through any surface bounded by the path changes over time.

It is important that you understand that the path may be real or purely imaginary – the magnetic field is produced whether the path exists or not.

Here's an expanded view of the Ampere–Maxwell law:

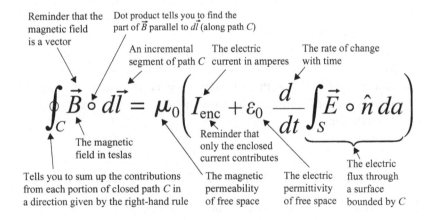

Of what use is the Ampere–Maxwell law? You can use it to determine the circulation of the magnetic field if you're given information about the enclosed current or the change in electric flux. Furthermore, in highly symmetric situations, you may be able to extract $\vec{B}$ from the dot product and the integral and determine the magnitude of the magnetic field.

$$\boxed{\oint_C \vec{B} \circ d\vec{l}} \text{ The magnetic field circulation}$$

Spend a few minutes moving a magnetic compass around a long, straight wire carrying a steady current, and here's what you're likely to find: the current in the wire produces a magnetic field that circles around the wire and gets weaker as you get farther from the wire.

With slightly more sophisticated equipment and an infinitely long wire, you'd find that the magnetic field strength decreases precisely as $1/r$, where r is the distance from the wire. So if you moved your measuring device in a way that kept the distance to the wire constant, say by circling around the wire as shown in Figure 4.1, the strength of the magnetic field wouldn't change. If you kept track of the direction of the magnetic field as you circled around the wire, you'd find that it always pointed along your path, perpendicular to an imaginary line from the wire to your location.

If you followed a random path around the wire getting closer and farther from the wire as you went around, you'd find the magnetic field getting stronger and weaker, and no longer pointing along your path.

Now imagine keeping track of the magnitude and direction of the magnetic field as you move around the wire in tiny increments. If, at each incremental step, you found the component of the magnetic field $\vec{B}$ along that portion of your path $d\vec{l}$, you'd be able to find $\vec{B} \circ d\vec{l}$. Keeping track of each value of $\vec{B} \circ d\vec{l}$ and then summing the results over your entire path, you'd have a discrete version of the left side of the Ampere–Maxwell law. Making this process continuous by letting the path increment shrink

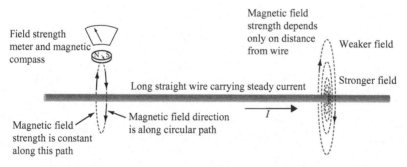

Figure 4.1 Exploring the magnetic field around a current-carrying wire.

toward zero would then give you the circulation of the magnetic field:

$$\text{Magnetic field circulation} = \oint_C \vec{B} \circ d\vec{l}. \qquad (4.1)$$

The Ampere–Maxwell law tells you that this quantity is proportional to the enclosed current and rate of change of electric flux through any surface bounded by your path of integration (C). But if you hope to use this law to determine the value of the magnetic field, you'll need to dig $\vec{B}$ out of the dot product and out of the integral. That means you'll have to choose your path around the wire very carefully – just as you had to choose a "special Gaussian surface" to extract the electric field from Gauss's law, you'll need a "special Amperian loop" to determine the magnetic field.

You'll find examples of how to do that after the next three sections, which discuss the terms on the right side of the Ampere–Maxwell law.

$\boxed{\mu_0}$ The permeability of free space

The constant of proportionality between the magnetic circulation on the left side of the Ampere–Maxwell law and the enclosed current and rate of flux change on the right side is μ_0, the permeability of free space. Just as the electric permittivity characterizes the response of a dielectric to an applied electric field, the magnetic permeability determines a material's response to an applied magnetic field. The permeability in the Ampere–Maxwell law is that of free space (or "vacuum permeability"), which is why it carries the subscript zero.

The value of the vacuum permeability in SI units is exactly $4\pi \times 10^{-7}$ volt-seconds per ampere-meter (Vs/Am); the units are sometimes given as newtons per square ampere (N/A^2) or the fundamental units of (m kg/C^2). Therefore, when you use the Ampere–Maxwell law, remember to multiply both terms on the right side by

$$\mu_0 = 4\pi \times 10^{-7} \text{ Vs/Am.}$$

As in the case of electric permittivity in Gauss's law for electric fields, the presence of this quantity does not mean that the Ampere–Maxwell law applies only to sources and fields in a vacuum. This form of the Ampere–Maxwell law is general, so long as you consider *all* currents (bound as well as free). In the Appendix, you'll find a version of this law that's more useful when dealing with currents and fields in magnetic materials.

One interesting difference between the effect of dielectrics on electric fields and the effect of magnetic substances on magnetic fields is that the magnetic field is actually *stronger* than the applied field within many magnetic materials. The reason for this is that these materials become magnetized when exposed to an external magnetic field, and the induced magnetic field is in the same direction as the applied field, as shown in Figure 4.2.

The permeability of a magnetic material is often expressed as the relative permeability, which is the factor by which the material's permeability exceeds that of free space:

$$\text{Relative permeability } \mu_r = \mu/\mu_0. \tag{4.2}$$

Materials are classified as diamagnetic, paramagnetic, or ferromagnetic on the basis of relative permeability. Diamagnetic materials have μ_r slightly less than 1.0 because the induced field weakly opposes the applied field. Examples of diamagnetic materials include gold and silver, which have μ_r of approximately 0.99997. The induced field within paramagnetic

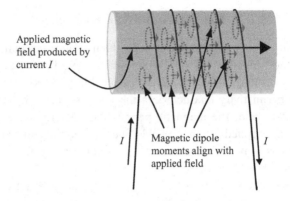

Figure 4.2 Effect of magnetic core on field inside solenoid.

materials weakly reinforces the applied field, so these materials have μ_r slightly greater than 1.0. One example of a paramagnetic material is aluminum with μ_r of 1.00002.

The situation is more complex for ferromagnetic materials, for which the permeability depends on the applied magnetic field. Typical maximum values of permeability range from several hundred for nickel and cobalt to over 5000 for reasonably pure iron.

As you may recall, the inductance of a long solenoid is given by the expression

$$L = \frac{\mu N^2 A}{\ell},$$
(4.3)

where μ is the magnetic permeability of the material within the solenoid, N is the number of turns, A is the cross-sectional area, and ℓ is the length of the coil. As this expression makes clear, adding an iron core to a solenoid may increase the inductance by a factor of 5000 or more.

Like electrical permittivity, the magnetic permeability of any medium is a fundamental parameter in the determination of the speed with which an electromagnetic wave propagates through that medium. This makes it possible to determine the speed of light in a vacuum simply by measuring μ_0 and ε_0 using an inductor and a capacitor; an experiment for which, to paraphrase Maxwell, the only use of light is to see the instruments.

$\boxed{I_{\text{enc}}}$ The enclosed electric current

Although the concept of "enclosed current" sounds simple, the question of exactly which current to include on the right side of the Ampere–Maxwell law requires careful consideration.

It should be clear from the first section of this chapter that the "enclosing" is done by the path C around which the magnetic field is integrated (if you're having trouble imagining a path enclosing anything, perhaps "encircling" is a better word). However, consider for a moment the paths and currents shown in Figure 4.3; which of the currents are enclosed by paths C_1, C_2, and C_3, and which are not?

The easiest way to answer that question is to imagine a membrane stretched across the path, as shown in Figure 4.4. The enclosed current is then just the net current that penetrates the membrane.

The reason for saying "net" current is that the direction of the current relative to the direction of integration must be considered. By convention, the right-hand rule determines whether a current is counted as positive or negative: if you wrap the fingers of your right hand around the path in the

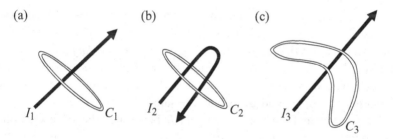

Figure 4.3 Currents enclosed (and not enclosed) by paths.

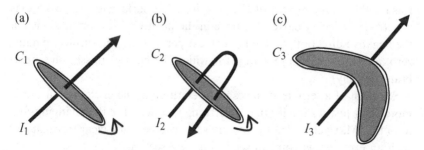

Figure 4.4 Membranes stretched across paths.

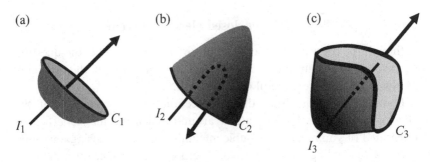

Figure 4.5 Alternative surfaces with boundaries C_1, C_2, and C_3.

direction of integration, your thumb points in the direction of positive current. Thus, the enclosed current in Figure 4.4(a) is $+I_1$ if the integration around path C_1 is performed in the direction indicated; it would be $-I_1$ if the integration were performed in the opposite direction.

Using the membrane approach and right-hand rule, you should be able to see that the enclosed current is zero in both Figure 4.4(b) and 4.4(c). No net current is enclosed in Figure 4.4(b) , since the sum of the currents is $I_2 + -I_2 = 0$, and no current penetrates the membrane in either direction in Figure 4.4(c).

An important concept for you to understand is that the enclosed current is exactly the same irrespective of the shape of the surface you choose, provided that the path of integration is a *boundary* (edge) of that surface. The surfaces shown in Figure 4.4 are the simplest, but you could equally well have chosen the surfaces shown in Figure 4.5, and the enclosed currents would be exactly the same.

Notice that in Figure 4.5(a) current I_1 penetrates the surface at only one point, so the enclosed current is $+I_1$, just as it was for the flat membrane of Figure 4.4(a). In Figure 4.5(b), current I_2 does not penetrate the "stocking cap" surface anywhere, so the enclosed current is zero, as it was for the flat membrane of Figure 4.4(b). The surface in Figure 4.5(c) is penetrated twice by current I_3, once in the positive direction and once in the negative direction, so the net current penetrating the surface remains zero, as it was in Figure 4.4(c) (for which the current missed the membrane entirely).

Selection of alternate surfaces and determining the enclosed current is more than just an intellectual diversion. The need for the changing-flux term that Maxwell added to Ampere's law can be made clear through just such an exercise, as you can see in the next section.

$$\boxed{\frac{d}{dt}\int_S \vec{E} \circ \hat{n}\, da}\ \text{The rate of change of flux}$$

This term is the electric flux analog of the changing magnetic flux term in Faraday's law, which you can read about in Chapter 3. In that case, a changing magnetic flux through any surface was found to induce a circulating electric field along a boundary path for that surface.

Purely by symmetry, you might suspect that a changing electric flux through a surface will induce a circulating magnetic field around a boundary of that surface. After all, magnetic fields are known to circulate – Ampere's law says that any electric current produces just such a circulating magnetic field. So how is it that several decades went by before anyone saw fit to write an "electric induction" law to go along with Faraday's law of magnetic induction?

For one thing, the magnetic fields induced by changing electric flux are extremely weak and are therefore very difficult to measure, so in the nineteenth century there was no experimental evidence on which to base such a law. In addition, symmetry is not always a reliable predictor between electricity and magnetism; the universe is rife with individual electric charges, but apparently devoid of the magnetic equivalent.

Maxwell and his contemporaries did realize that Ampere's law as originally conceived applies only to steady electric currents, since it is consistent with the principle of conservation of charge only under static conditions. To better understand the relationship between magnetic fields and electric currents, Maxwell worked out an elaborate conceptual model in which magnetic fields were represented by mechanical vortices and electric currents by the motion of small particles pushed along by the whirling vortices. When he added elasticity to his model and allowed the magnetic vortices to deform under stress, Maxwell came to understand the need for an additional term in his mechanical version of Ampere's law. With that understanding, Maxwell was able to discard his mechanical model and rewrite Ampere's law with an additional source for magnetic fields. That source is the changing electric flux in the Ampere–Maxwell law.

Most texts use one of three approaches to demonstrating the need for the changing-flux term in the Ampere–Maxwell law: conservation of charge, special relativity, or an inconsistency in Ampere's law when applied to a charging capacitor. This last approach is the most common, and is the one explained in this section.

Consider the circuit shown in Figure 4.6. When the switch is closed, a current I flows as the battery charges the capacitor. This current produces

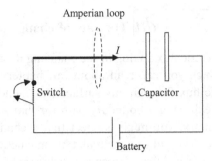

Figure 4.6 Charging capacitor.

a magnetic field around the wires, and the circulation of that field is given by Ampere's law

$$\oint_C \vec{B} \circ d\vec{l} = \mu_0(I_{\text{enc}}).$$

A serious problem arises in determining the enclosed current. According to Ampere's law, the enclosed current includes all currents that penetrate any surface for which path C is a boundary. However, you'll get very different answers for the enclosed current if you choose a flat membrane as your surface, as shown in Figure 4.7(a), or a "stocking cap" surface as shown in Figure 4.7(b).

Although current I penetrates the flat membrane as the capacitor charges, no current penetrates the "stocking cap" surface (since the charge accumulates at the capacitor plate). Yet the Amperian loop is a boundary to both surfaces, and the integral of the magnetic field around that loop must be same no matter which surface you choose.

You should note that this inconsistency occurs only while the capacitor is charging. No current flows before the switch is thrown, and after the capacitor is fully charged the current returns to zero. In both of these circumstances, the enclosed current is zero through any surface you can imagine. Therefore, any revision to Ampere's law must retain its correct behavior in static situations while extending its utility to charging capacitors and other time-dependent situations.

With more than a little hindsight, we might phrase our question this way: since no conduction current flows between the capacitor plates, what else might be going on in that region that would serve as the source of a magnetic field?

Since charge is accumulating on the plates as the capacitor charges up, you know that the electric field between the plates must be changing with

(a)

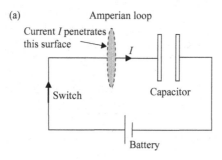

(b)

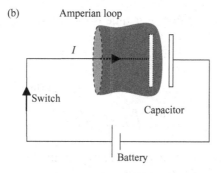

Figure 4.7 Alternative surfaces for determining enclosed current.

time. This means that the electric flux through the portion of your "stocking cap" surface between the plates must also be changing, and you can use Gauss's law for electric fields to determine the change in flux.

By shaping your surface carefully, as in Figure 4.8, you can make it into a "special Gaussian surface", which is everywhere perpendicular to the electric field and over which the electric field is either uniform or zero. Neglecting edge effects, the electric field between two charged conducting plates is $\vec{E} = (\sigma/\varepsilon_0)\,\hat{n}$, where σ is the charge density on the plates (Q/A), making the electric flux through the surface

$$\Phi_E = \int_S \vec{E} \circ \hat{n}\, da = \int_S \frac{\sigma}{\varepsilon_0} da = \frac{Q}{A\varepsilon_0} \int_S da = \frac{Q}{\varepsilon_0}. \qquad (4.4)$$

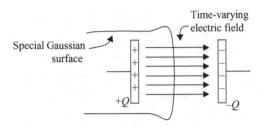

Figure 4.8 Changing electric flux between capacitor plates.

The change in electric flux over time is therefore,

$$\frac{d}{dt}\left(\int_S \vec{E} \circ \hat{n}\, da\right) = \frac{d}{dt}\left(\frac{Q}{\varepsilon_0}\right) = \frac{1}{\varepsilon_0}\frac{dQ}{dt}. \qquad (4.5)$$

Multiplying by the vacuum permittivity makes this

$$\varepsilon_0 \frac{d}{dt}\left(\int_S \vec{E} \circ \hat{n}\, da\right) = \frac{dQ}{dt}. \qquad (4.6)$$

Thus, the change in electric flux with time multiplied by permittivity has units of charge divided by time (coulombs per second or amperes in SI units), which are of course the units of *current*. Moreover a current-like quantity is exactly what you might expect to be the additional source of the magnetic field around your surface boundary. For historical reasons, the product of the permittivity and the change of electric flux through a surface is called the "displacement current" even though no charge actually flows across the surface. The displacement current is defined by the relation

$$I_d \equiv \varepsilon_0 \frac{d}{dt}\left(\int_S \vec{E} \circ \hat{n}\, da\right). \qquad (4.7)$$

Whatever you choose to call it, Maxwell's addition of this term to Ampere's law demonstrated his deep physical insight and set the stage for his subsequent discovery of the electromagnetic nature of light.

$$\oint_C \vec{B} \circ d\vec{l} = \mu_0\left(I_{\text{enc}} + \varepsilon_0 \frac{d}{dt}\int_S \vec{E} \circ \hat{n}\, da\right)$$ **Applying the Ampere–Maxwell law (integral form)**

Like the electric field in Gauss's law, the magnetic field in the Ampere–Maxwell law is buried within an integral and coupled to another vector quantity by the dot product. As you might expect, it is only in highly symmetric situations that you'll be able to determine the magnetic field using this law. Fortunately, several interesting and realistic geometries possess the requisite symmetry, including long current-carrying wires and parallel-plate capacitors.

For such problems, the challenge is to find an Amperian loop over which you expect $\vec{B}$ to be uniform and at a constant angle to the loop. However, how do you know what to expect for $\vec{B}$ before you solve the problem?

In many cases, you'll already have some idea of the behavior of the magnetic field on the basis of your past experience or from experimental evidence. What if that's not the case – how are you supposed to figure out how to draw your Amperian loop?

There's no single answer to that question, but the best approach is to use logic to try to reason your way to a useful result. Even for complex geometries, you may be able to use the Biot–Savart law to discern the field direction by eliminating some of the components through symmetry considerations. Alternatively, you can imagine various behaviors for $\vec{B}$ and then see if they lead to sensible consequences.

For example, in a problem involving a long, straight wire, you might reason as follows: the magnitude of $\vec{B}$ must get smaller as you move away from the wire; otherwise Oersted's demonstration in Denmark would have deflected compass needles everywhere in the world, which it clearly did not. Furthermore, since the wire is round, there's no reason to expect that the magnetic field on one side of the wire is any different from the field on the other side. So if $\vec{B}$ decreases with distance from the wire and is the same all around the wire, you can safely conclude that one *path of constant $\vec{B}$* would be a circle centered on the wire and perpendicular to the direction of current flow.

However, to deal with the dot product of $\vec{B}$ and $d\vec{l}$ in Ampere's law, you also need to make sure that your path maintains a constant angle (preferably 0°) to the magnetic field. If $\vec{B}$ were to have both radial and transverse components that vary with distance, the angle between your path and the magnetic field might depend on distance from the wire.

If you understand the cross–product between $d\vec{l}$ and $\hat{r}$ in the Biot–Savart law, you probably suspect that this is not the case. To verify that, imagine that $\vec{B}$ has a component pointing directly toward the wire. If you were to look along the wire in the direction of the current, you'd see the current running away from you and the magnetic field pointing at the wire. Moreover, if you had a friend looking in the opposite direction at the same time, she'd see the current coming toward her, and of course she would also see $\vec{B}$ pointing toward the wire.

Now ask yourself, what would happen if you reversed the direction of the current flow. Since the magnetic field is linearly proportional to the current $(\vec{B} \propto \vec{I})$ according to the Biot–Savart law, reversing the current must also reverse the magnetic field, and $\vec{B}$ would then point away from the wire. Now looking in your original direction, you'd see a current coming toward you (since it was going away from you before it was reversed), but now you'd see the magnetic field pointing away from the wire. Moreover, your friend, still looking in her original direction, would see the current running away from her, but with the magnetic field pointing away from the wire.

Comparing notes with your friend, you'd find a logical inconsistency. You'd say, "currents traveling away from me produce a magnetic field pointing toward the wire, and currents coming toward me produce a magnetic field pointing away from the wire." Your friend, of course, would report exactly the opposite behavior. In addition, if you switched positions and repeated the experiment, you'd each find that your original conclusions were no longer true.

This inconsistency is resolved if the magnetic field circles around the wire, having no radial component at all. With $\vec{B}$ having only a φ-component,[5] all observers agree that currents traveling away from an observer produce clockwise magnetic fields as seen by that observer, whereas currents approaching an observer produce counterclockwise magnetic fields for that observer.

In the absence of external evidence, this kind of logical reasoning is your best guide to designing useful Amperian loops. Therefore, for problems involving a straight wire, the logical choice for your loop is a circle centered on the wire. How big should you make your loop? Remember why you're making an Amperian loop in the first place – to find the value of the magnetic field at some location. So *make your*

---

[5] Remember that there's a review of cylindrical and spherical coordinates on the book's website.

Amperian loop go through that location. In other words, the loop radius should be equal to the distance from the wire at which you intend to find the value of the magnetic field. The following example shows how this works.

Example 4.1: Given the current in a wire, find the magnetic field within and outside the wire.

Problem: A long, straight wire of radius r_0 carries a steady current I uniformly distributed throughout its cross-sectional area. Find the magnitude of the magnetic field as a function of r, where r is the distance from the center of the wire, for both $r > r_0$ and $r < r_0$.

Solution: Since the current is steady, you can use Ampere's law in its original form

$$\oint_C \vec{B} \circ d\vec{l} = \mu_0(I_{enc}).$$

To find $\vec{B}$ at exterior points ($r>r_0$), use the logic described above and draw your loop outside the wire, as shown by Amperian loop #1 in Figure 4.9. Since both $\vec{B}$ and $d\vec{l}$ have only φ-components and point in the same direction if you obey the right-hand rule in determining your direction of integration, the dot product $\vec{B} \circ d\vec{l}$ becomes $|\vec{B}|\,|d\vec{l}|\cos(0°)$. Furthermore, since $|\vec{B}|$ is constant around your loop, it comes out of the integral:

$$\oint_C \vec{B} \circ d\vec{l} = \oint_C |\vec{B}|\,|d\vec{l}| = B \oint_C dl = B(2\pi r),$$

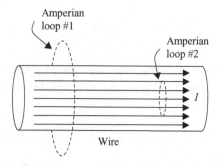

Figure 4.9 Amperian loops for current-carrying wire of radius r_0.

where r is the radius of your Amperian loop.[6] Ampere's law tells you that the integral of $\vec{B}$ around your loop is equal to the enclosed current times the permeability of free space, and the enclosed current in this case is all of I, so

$$B(2\pi r) = \mu_0 I_{enc} = \mu_0 I$$

and, since $\vec{B}$ is in the φ-direction,

$$\vec{B} = \frac{\mu_0 I}{2\pi r}\hat{\varphi},$$

as given in Table 2.1. Note that this means that at points outside the wire the magnetic field decreases as $1/r$ and behaves as if all the current were at the center of the wire.

To find the magnetic field within the wire, you can apply the same logic and use a smaller loop, as shown by Amperian loop #2 in Figure 4.9. The only difference in this case is that not all the current is enclosed by the loop; since the current is distributed uniformly throughout the wire's cross section, the current density[7] is $I/(\pi r_0^2)$, and the current passing through the loop is simply that density times the area of the loop. Thus,

$$\text{Enclosed current} = \text{current density} \times \text{loop area}$$

or

$$I_{enc} = \frac{I}{\pi r_0^2}\pi r^2 = I\frac{r^2}{r_0^2}.$$

Inserting this into Ampere's law gives

$$\oint_C \vec{B} \circ d\vec{l} = B(2\pi r) = \mu_0 I_{enc} = \mu_0 I\frac{r^2}{r_0^2},$$

or

$$B = \frac{\mu_0 I r}{2\pi r_0^2}.$$

Thus, inside the wire the magnetic field increases linearly with distance from the center of the wire, reaching a maximum at the surface of the wire.

---

[6] Another way to understand this is to write $\vec{B}$ as $B_\varphi \hat{\varphi}$ and $d\vec{l}$ as $(rd\varphi)\,\hat{\varphi}$, so $\vec{B} \circ d\vec{l} = B_\varphi rd\varphi$ and $\int_0^{2\pi} B_\varphi rd\varphi = B_\varphi(2\pi r)$.

[7] If you need a review of current density, you'll find a section covering this topic later in this chapter.

Example 4.2: Given the time-dependent charge on a capacitor, find the rate of change of the electric flux between the plates and the magnitude of the resulting magnetic field at a specified location.

Problem: A battery with potential difference ΔV charges a circular parallel-plate capacitor of capacitance C and plate radius r_0 through a wire with resistance R. Find the rate of change of the electric flux between the plates as a function of time and the magnetic field at a distance r from the center of the plates.

Solution: From Equation 4.5, the rate of change of electric flux between the plates is

$$\frac{d\Phi_E}{dt} = \frac{d}{dt}\left(\int_S \vec{E} \circ \hat{n}\, da \right) = \frac{1}{\varepsilon_0}\frac{dQ}{dt},$$

where Q is the total charge on each plate. So you should begin by determining how the charge on a capacitor plate changes with time as the capacitor is charged. If you've studied series RC circuits, you may recall that the relevant expression is

$$Q(t) = C\Delta V(1 - e^{-t/RC}),$$

where ΔV, R, and C represent the potential difference, the series resistance, and the capacitance, respectively. Thus,

$$\frac{d\Phi_E}{dt} = \frac{1}{\varepsilon_0}\frac{d}{dt}\left[C\Delta V(1 - e^{-t/RC}) \right] = \frac{1}{\varepsilon_0}\left(C\Delta V \frac{1}{RC} e^{-t/RC} \right) = \frac{\Delta V}{\varepsilon_0 R} e^{-t/RC}.$$

This is the rate of change of the total electric flux between the plates. To find the magnetic field at a distance r from the center of the plates, you're going to have to construct a special Amperian loop to help you extract the magnetic field from the integral in the Ampere–Maxwell law:

$$\oint_C \vec{B} \circ d\vec{l} = \mu_0\left(I_{\text{enc}} + \varepsilon_0 \frac{d}{dt}\int_S \vec{E} \circ \hat{n}\, da \right).$$

Since no charge flows between the capacitor plates, $I_{\text{enc}} = 0$, and

$$\oint_C \vec{B} \circ d\vec{l} = \mu_0\left(\varepsilon_0 \frac{d}{dt}\int_S \vec{E} \circ \hat{n}\, da \right).$$

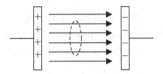

Figure 4.10 Amperian loop between capacitor plates.

As in the previous example, you're faced with the challenge of designing a special Amperian loop around which the magnetic field is constant in amplitude and parallel to the path of integration around the loop. If you use similar logic to that for the straight wire, you'll see that the best choice is to make a loop parallel to the plates, as shown in Figure 4.10.

The radius of this loop is r, the distance from the center of the plates at which you are trying to find the magnetic field. Of course, not all of the flux between the plates passes through this loop, so you will have to modify your expression for the flux change accordingly. The fraction of the total flux that passes through a loop of radius r is simply the ratio of the loop area to the capacitor plate area, which is $\pi r^2 / \pi r_0^2$, so the rate of change of flux through the loop is

$$\left(\frac{d\Phi_E}{dt}\right)_{\text{Loop}} = \frac{\Delta V}{\varepsilon_0 R} e^{-t/RC} \left(\frac{r^2}{r_0^2}\right).$$

Inserting this into the Ampere–Maxwell law gives

$$\oint_C \vec{B} \circ d\vec{l} = \mu_0 \left[\varepsilon_0 \frac{\Delta V}{\varepsilon_0 R} e^{-t/RC} \left(\frac{r^2}{r_0^2}\right)\right] = \frac{\mu_0 \Delta V}{R} e^{-t/RC} \left(\frac{r^2}{r_0^2}\right).$$

Moreover, since you've chosen your Amperian loop so as to allow $\vec{B}$ to come out of the dot product and the integral using the same symmetry arguments as in Example 4.1,

$$\oint_C \vec{B} \circ d\vec{l} = B(2\pi r) = \frac{\mu_0 \Delta V}{R} e^{-t/RC} \left(\frac{r^2}{r_0^2}\right),$$

which gives

$$B = \frac{\mu_0 \Delta V}{2\pi r R} e^{-t/RC} \left(\frac{r^2}{r_0^2}\right) = \frac{\mu_0 \Delta V}{2\pi R} e^{-t/RC} \left(\frac{r}{r_0^2}\right),$$

meaning that the magnetic field increases linearly with distance from the center of the capacitor plates and decreases exponentially with time, reaching $1/e$ of its original value at time $t = RC$.

4.2 The differential form of the Ampere–Maxwell law

The differential form of the Ampere–Maxwell law is generally written as

$$\vec{\nabla} \times \vec{B} = \mu_0 \left(\vec{J} + \varepsilon_0 \frac{\partial \vec{E}}{\partial t} \right) \qquad \text{The Ampere–Maxwell law.}$$

The left side of this equation is a mathematical description of the curl of the magnetic field – the tendency of the field lines to circulate around a point. The two terms on the right side represent the electric current density and the time rate of change of the electric field.

These terms are discussed in detail in the following sections. For now, make sure you grasp the main idea of the differential form of the Ampere–Maxwell law:

A circulating magnetic field is produced by an electric current and by an electric field that changes with time.

To help you understand the meaning of each symbol in the differential form of the Ampere–Maxwell law, here's an expanded view:

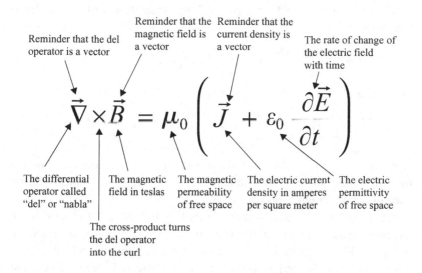

$\boxed{\vec{\nabla} \times \vec{B}}$ The curl of the magnetic field

The left side of the differential form of the Ampere–Maxwell law represents the curl of the magnetic field. All magnetic fields, whether produced by electrical currents or by changing electric fields, circulate back upon themselves and form continuous loops. In addition all fields that circulate back on themselves must include at least one location about which the path integral of the field is nonzero. For the magnetic field, locations of nonzero curl are locations at which current is flowing or an electric field is changing.

It is important that you understand that just because magnetic fields circulate, you should not conclude that the curl is nonzero everywhere in the field. A common misconception is that the curl of a vector field is nonzero wherever the field appears to curve.

To understand why that is not correct, consider the magnetic field of the infinite line current shown in Figure 2.1. The magnetic field lines circulate around the current, and you know from Table 2.1 that the magnetic field points in the $\hat{\varphi}$ direction and decreases as $1/r$

$$\vec{B} = \frac{\mu_0 I}{2\pi r} \hat{\varphi}.$$

Finding the curl of this field is particularly straightforward in cylindrical coordinates

$$\vec{\nabla} \times \vec{B} = \left(\frac{1}{r} \frac{\partial B_z}{\partial \varphi} - \frac{\partial B_\varphi}{\partial z} \right) \hat{r} + \left(\frac{\partial B_r}{\partial z} - \frac{\partial B_z}{\partial r} \right) \hat{\varphi} + \frac{1}{r} \left(\frac{\partial (r B_\varphi)}{\partial r} - \frac{\partial B_r}{\partial \varphi} \right) \hat{z}.$$

Since B_r and B_z are both zero, this is

$$\vec{\nabla} \times \vec{B} = \left(-\frac{\partial B_\varphi}{\partial z} \right) \hat{r} + \frac{1}{r} \left(\frac{\partial (r B_\varphi)}{\partial r} \right) \hat{z} = -\frac{\partial (\mu_0 I / 2\pi r)}{\partial z} \hat{r} + \frac{1}{r} \frac{\partial (r \mu_0 I / 2\pi r)}{\partial r} \hat{z} = 0.$$

However, doesn't the differential form of the Ampere–Maxwell law tell us that the curl of the magnetic field is nonzero in the vicinity of electric currents and changing electric fields?

No, it doesn't. It tells us that the curl of $\vec{B}$ is nonzero *exactly at the location* through which an electric current is flowing, or at which an electric field is changing. Away from that location, the field definitely does curve, but the curl at any given point is precisely zero, as you just found from the equation for the magnetic field of an infinite line current.

How can a curving field have zero curl? The answer lies in the *amplitude* as well as the direction of the magnetic field, as you can see in Figure 4.11.

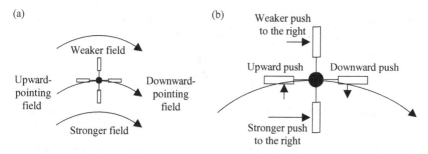

Figure 4.11 Offsetting components of the curl of $\vec{B}$.

Using the fluid-flow and small paddlewheel analogy, imagine the forces on the paddlewheel placed in the field shown in Figure 4.11(a). The center of curvature is well below the bottom of the figure, and the spacing of the arrows indicates that the field is getting weaker with distance from the center. At first glance, it may seem that this paddlewheel would rotate clockwise owing to the curvature of the field, since the flow lines are pointing slightly upward at the left paddle and slightly downward at the right. However, consider the effect of the weakening of the field above the axis of the paddlewheel: the top paddle receives a weaker push from the field than the bottom paddle, as shown in Figure 4.11(b). The stronger force on the bottom paddle will attempt to cause the paddlewheel to rotate counterclockwise. Thus, the downward curvature of the field is offset by the weakening of the field with distance from the center of curvature. And if the field diminishes as $1/r$, the upward–downward push on the left and right paddles is exactly compensated by the weaker–stronger push on the top and bottom paddles. The clockwise and counter-clockwise forces balance, and the paddlewheel does not turn – the curl at this location is zero, even though the field lines are curved.

The key concept in this explanation is that the magnetic field may be curved at many different locations, but only at points at which current is flowing (or the electric flux is changing) is the curl of $\vec{B}$ nonzero. This is analogous to the $1/r^2$ reduction in electric field amplitude with distance from a point charge, which keeps the divergence of the electric field as zero at all points away from the location of the charge.

As in the electric field case, the reason the origin (where $r=0$) is not included in our previous analysis is that our expression for the curl includes terms containing r in the denominator, and those terms become infinite at the origin. To evaluate the curl at the origin, use the formal definition of curl as described in Chapter 3:

$$\vec{\nabla} \times \vec{B} \equiv \lim_{\Delta S \to 0} \frac{1}{\Delta S} \oint_C \vec{B} \circ \vec{dl}$$

Considering a special Amperian loop surrounding the current, this is

$$\vec{\nabla} \times \vec{B} \equiv \lim_{\Delta S \to 0} \frac{1}{\Delta S} \oint_C \vec{B} \circ \vec{dl} = \lim_{\Delta S \to 0} \left(\frac{1}{\Delta S} \frac{\mu_0 \vec{I}}{2\pi r} (2\pi r) \right) = \lim_{\Delta S \to 0} \left(\frac{1}{\Delta S} \mu_0 \vec{I} \right).$$

However, $\vec{I}/\Delta S$ is just the average current density over the surface ΔS, and as ΔS shrinks to zero, this becomes equal to $\vec{J}$, the current density at the origin. Thus, at the origin

$$\vec{\nabla} \times \vec{B} = \mu_0 \vec{J}$$

in accordance with Ampere's law.

So just as you might be fooled into thinking that charge-based electric field vectors "diverge" everywhere because they get farther apart, you might also think that magnetic field vectors have curl everywhere because they curve around a central point. But the key factor in determining the curl at any point is not simply the curvature of the field lines at that point, but how the change in the field from one side of the point to the other (say from left to right) compares to the change in the field in the orthogonal direction (below to above). If those spatial derivatives are precisely equal, then the curl is zero at that point.

In the case of a current-carrying wire, the reduction in the amplitude of the magnetic field away from the wire exactly compensates for the curvature of the field lines. Thus, the curl of the magnetic field is zero everywhere except at the wire itself, where electric current is flowing.

$\boxed{\vec{J}}$ The electric current density

The right side of the differential form of the Ampere–Maxwell law contains two source terms for the circulating magnetic field; the first involves the vector electric current density. This is sometimes called the "volume current density," which can be a source of confusion if you're accustomed to "volume density" meaning the amount of something per unit volume, such as kg/m^3 for mass density or C/m^3 for charge density.

This is not the case for current density, which is defined as the vector current flowing through a unit cross-sectional area perpendicular to the direction of the current. Thus, the units of current density are not amperes per cubic meter, but rather amperes per square meter (A/m^2).

To understand the concept of current density, recall that in the discussion of flux in Chapter 1, the quantity $\vec{A}$ is defined as the number density of the fluid (particles per cubic meter) times the velocity of the flow (meters per second). As the product of the number density (a scalar) and the velocity (a vector), $\vec{A}$ is a vector in the same direction as the velocity, with units of particles per square meter per second. To find the number of particles per second passing through a surface in the simplest case ($\vec{A}$ uniform and perpendicular to the surface), you simply multiply $\vec{A}$ by the area of the surface.

These same concepts are relevant for current density, provided we consider the *amount of charge* passing through the surface rather than the number of atoms. If the number density of charge carriers is n and the charge per carrier is q, then the amount of charge passing through a unit area perpendicular to the flow per second is

$$\vec{J} = nq\vec{v}_d \quad (C/m^2 \text{ s or } A/m^2), \tag{4.8}$$

where $\vec{v}_d$ is the average drift velocity of charge carriers. Thus, the direction of the current density is the direction of current flow, and the magnitude is the current per unit area, as shown in Figure 4.12.

The complexity of the relationship between the total current I through a surface and the current density $\vec{J}$ depends on the geometry of the situation. If the current density $\vec{J}$ is uniform over a surface S and is everywhere perpendicular to the surface, the relationship is

$$I = |\vec{J}| \times (\text{surface area}) \quad \vec{J} \text{ uniform and perpendicular to } S. \tag{4.9}$$

A student's guide to Maxwell's Equations

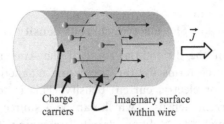

Figure 4.12 Charge flow and current density.

If $\vec{J}$ is uniform over a surface S but is not necessarily perpendicular to the surface, to find the total current I through S you'll have to determine the component of the current density perpendicular to the surface. This makes the relationship between I and $\vec{J}$:

$$I = \vec{J} \circ \hat{n} \times (\text{surface area}) \quad \vec{J} \text{ uniform and at an angle to } S. \quad (4.10)$$

And, if $\vec{J}$ is nonuniform and not perpendicular to the surface, then

$$I = \int_{S} \vec{J} \circ \hat{n} \, da \quad \vec{J} \text{ nonuniform and at a variable angle to } S. \quad (4.11)$$

This expression explains why some texts refer to electric current as "the flux of the current density."

The electric current density in the Ampere–Maxwell law includes all currents, including the bound current density in magnetic materials. You can read more about Maxwell's Equations inside matter in the Appendix.

$\boxed{\varepsilon_0 \dfrac{\partial \vec{E}}{\partial t}}$ The displacement current density

The second source term for the magnetic field in the Ampere–Maxwell law involves the rate of change of the electric field with time. When multiplied by the electrical permittivity of free space, this term has SI units of amperes per square meter. These units are identical to those of $\vec{J}$, the conduction current density that also appears on the right side of the differential form of the Ampere–Maxwell law. Maxwell originally attributed this term to the physical displacement of electrical particles caused by elastic deformation of magnetic vortices, and others coined the term "displacement current" to describe the effect.

However, does the displacement current density represent an actual current? Certainly not in the conventional sense of the word, since electric current is defined as the physical movement of charge. But it is easy to understand why a term that has units of amperes per square meter and acts as a source of the magnetic field has retained that name over the years. Furthermore, the displacement current density is a vector quantity that has the same relationship to the magnetic field as does $\vec{J}$, the conduction current density.

The key concept here is that a changing electric field produces a changing magnetic field even when no charges are present and no physical current flows. Through this mechanism, electromagnetic waves may propagate through even a perfect vacuum, as changing magnetic fields induce electric fields, and changing electric fields induce magnetic fields.

The importance of the displacement current term, which arose initially from Maxwell's mechanical model, is difficult to overstate. Adding a changing electric field as a source of the magnetic field certainly extended the scope of Ampere's law to time-dependent fields by eliminating the inconsistency with the principle of conservation of charge. Far more importantly, it allowed James Clerk Maxwell to establish a comprehensive theory of electromagnetism, the first true field theory and the foundation for much of twentieth century physics.

$$\boxed{\vec{\nabla} \times \vec{B} = \mu_0 \left(\vec{J} + \varepsilon_0 \frac{\partial \vec{E}}{\partial t} \right)}$$ **Applying the Ampere–Maxwell law (differential form)**

The most common applications of the differential form of the Ampere–Maxwell law are problems in which you're provided with an expression for the vector magnetic field and you're asked to determine the electric current density or the displacement current. Here are two examples of this kind of problem.

Example 4.3: Given the magnetic field, find the current density at a specified location.

Problem: Use the expressions for the magnetic field in Table 2.1 to find the current density both inside and outside a long, straight wire of radius r_0 carrying current I uniformly throughout its volume in the positive z-direction.

Solution: From Table 2.1 and Example 4.1, the magnetic field inside a long, straight wire is

$$\vec{B} = \frac{\mu_0 I r}{2\pi r_0^2} \hat{\varphi},$$

where I is the current in the wire and r_0 is the wire's radius. In cylindrical coordinates, the curl of $\vec{B}$ is

$$\vec{\nabla} \times \vec{B} \equiv \left(\frac{1}{r} \frac{\partial B_z}{\partial \varphi} - \frac{\partial B_\varphi}{\partial z} \right) \hat{r} + \left(\frac{\partial B_r}{\partial z} - \frac{\partial B_z}{\partial r} \right) \hat{\varphi} + \frac{1}{r} \left(\frac{\partial (r B_\varphi)}{\partial r} - \frac{\partial B_r}{\partial \varphi} \right) \hat{z}.$$

And, since $\vec{B}$ has only a $\hat{\varphi}$-component in this case,

$$\vec{\nabla} \times \vec{B} = \left(-\frac{\partial B_\varphi}{\partial z} \right) \hat{r} + \frac{1}{r} \left(\frac{\partial (r B_\varphi)}{\partial r} \right) \hat{z} = \frac{1}{r} \left(\frac{\partial \left(r (\mu_0 I r / 2\pi r_0^2) \right)}{\partial r} \right) \hat{z}$$

$$= \frac{1}{r} \left(2r \frac{\mu_0 I}{2\pi r_0^2} \right) \hat{z} = \left(\frac{\mu_0 I}{\pi r_0^2} \right) \hat{z}.$$

Using the static version of the Ampere–Maxwell law (since the current is steady), you can find $\vec{J}$ from the curl of $\vec{B}$:

$$\vec{\nabla} \times \vec{B} = \mu_0 (\vec{J}).$$

Thus,

$$\vec{J} = \frac{1}{\mu_0} \left(\frac{\mu_0 I}{\pi r_0^2} \right) \hat{z} = \frac{I}{\pi r_0^2} \hat{z},$$

which is the current density within the wire. Taking the curl of the expression for $\vec{B}$ outside the wire, you'll find that $\vec{J} = 0$, as expected.

Example 4.4: Given the magnetic field, find the displacement current density.

Problem: The expression for the magnetic field of a circular parallel-plate capacitor found in Example 4.2 is

$$\vec{B} = \frac{\mu_0 \Delta V}{2\pi R} e^{-t/RC} \left(\frac{r}{r_0^2} \right) \hat{\varphi}.$$

Use this result to find the displacement current density between the plates.

Solution: Once again you can use the curl of $\vec{B}$ in cylindrical coordinates:

$$\vec{\nabla} \times \vec{B} \equiv \left(\frac{1}{r} \frac{\partial B_z}{\partial \varphi} - \frac{\partial B_\varphi}{\partial z} \right) \hat{r} + \left(\frac{\partial B_r}{\partial z} - \frac{\partial B_z}{\partial r} \right) \hat{\varphi} + \frac{1}{r} \left(\frac{\partial (r B_\varphi)}{\partial r} - \frac{\partial B_r}{\partial \varphi} \right) \hat{z}.$$

And, once again $\vec{B}$ has only a $\hat{\varphi}$-component:

$$\vec{\nabla} \times \vec{B} = \left(-\frac{\partial B_\varphi}{\partial z} \right) \hat{r} + \frac{1}{r} \left(\frac{\partial (r B_\varphi)}{\partial r} \right) \hat{z} = \frac{1}{r} \left[\frac{\partial \left((r \mu_0 \Delta V / 2\pi R) e^{-t/RC} \left(\frac{r}{r_0^2} \right) \right)}{\partial r} \right] \hat{z}$$

$$= \frac{1}{r} \left[2r \frac{\mu_0 \Delta V}{2\pi R} e^{-t/RC} \left(\frac{1}{r_0^2} \right) \right] \hat{z} = \left[\frac{\mu_0 \Delta V}{\pi R} e^{-t/RC} \left(\frac{1}{r_0^2} \right) \right] \hat{z}.$$

Since there is no conduction current between the plates, $\vec{J} = 0$ in this case and the Ampere–Maxwell law is

$$\vec{\nabla} \times \vec{B} = \mu_0 \left(\varepsilon_0 \frac{\partial \vec{E}}{\partial t} \right),$$

from which you can find the displacement current density,

$$\varepsilon_0 \frac{\partial \vec{E}}{\partial t} = \frac{\vec{\nabla} \times \vec{B}}{\mu_0} = \frac{1}{\mu_0} \left[\frac{\mu_0 \Delta V}{\pi R} e^{-t/RC} \left(\frac{1}{r_0^2} \right) \right] \hat{z} = \left[\frac{\Delta V}{R} e^{-t/RC} \frac{1}{\pi r_0^2} \right] \hat{z}.$$

Problems

The following problems will test your understanding of the Ampere–Maxwell law. Full solutions are available on the book's website.

4.1 Two parallel wires carry currents I_1 and $2I_1$ in opposite directions. Use Ampere's law to find the magnetic field at a point midway between the wires.

4.2 Find the magnetic field inside a solenoid (Hint: use the Amperian loop shown in the figure, and use the fact that the field is parallel to the axis of the solenoid and negligible outside).

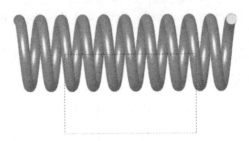

4.3 Use the Amperian loop shown in the figure to find the magnetic field within a torus.

4.4 The coaxial cable shown in the figure carries current I_1 in the direction shown on the inner conductor and current I_2 in the opposite direction on the outer conductor. Find the magnetic field in the space between the conductors as well as outside the cable if the magnitudes of I_1 and I_2 are equal.

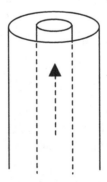

4.5 Find the displacement current produced between the plates of a discharging capacitor for which the charge varies as

$$Q(t) = Q_0 e^{-t/RC},$$

where Q_0 is the initial charge, C is the capacitance of the capacitor, and R is the resistance of the circuit through which the capacitor is discharging.

4.6 A magnetic field of $\vec{B} = a\sin(by)e^{bx}\hat{z}$ is produced by an electric current. What is the density of that current?

4.7 Find the electric current density that produces a magnetic field given by $\vec{B} = B_0(e^{-2r}\sin\varphi)\hat{z}$ in cylindrical coordinates.

4.8 What density of current would produce a magnetic field given by $\vec{B} = (a/r + b/re^{-r} + ce^{-r})\hat{\varphi}$ in cylindrical coordinates?

4.9 In this chapter, you learned that the magnetic field of a long, straight wire, given by

$$\vec{B} = \frac{\mu_0 I}{2\pi r}\hat{\varphi},$$

has zero curl everywhere except at the wire itself. Prove that this would not be true for a field that decreases as $1/r^2$ with distance.

4.10 To directly measure the displacement current, researchers use a time-varying voltage to charge and discharge a circular parallel-plate capacitor. Find the displacement current density and electric field as a function of time that would produce a magnetic field given by

$$\vec{B} = \frac{r\omega\,\Delta V\cos(\omega t)}{2d(c^2)}\hat{\varphi},$$

where r is the distance from the center of the capacitor, ω is the angular frequency of the applied voltage ΔV, d is the plate spacing, and c is the speed of light.

5

From Maxwell's Equations to the wave equation

Each of the four equations that have come to be known as Maxwell's Equations is powerful in its own right, for each embodies an important aspect of electromagnetic field theory. However, Maxwell's achievement went beyond the synthesis of these laws or the addition of the displacement current term to Ampere's law – it was by considering these equations *in combination* that he reached his goal of developing a comprehensive theory of electromagnetism. That theory elucidated the true nature of light itself and opened the eyes of the world to the full spectrum of electromagnetic radiation.

In this chapter, you'll learn how Maxwell's Equations lead directly to the wave equation in just a few steps. To make those steps, you'll first have to understand two important theorems of vector calculus: the divergence theorem and Stokes' theorem. These two theorems make the transition from the integral form to the differential form of Maxwell's Equations quite straightforward:

Gauss's law for electric fields:

$$\oint_S \vec{E} \circ \hat{n}\, da = q_{\text{in}}/\varepsilon_0 \xrightarrow[\text{theorem}]{\text{Divergence}} \vec{\nabla} \circ \vec{E} = \rho/\varepsilon_0.$$

Gauss's law for magnetic fields:

$$\oint_S \vec{B} \circ \hat{n}\, da = 0 \xrightarrow[\text{theorem}]{\text{Divergence}} \vec{\nabla} \circ \vec{B} = 0.$$

Faraday's law:

$$\oint_C \vec{E} \circ d\vec{l} = -\frac{d}{dt}\int_S \vec{B} \circ \hat{n}\, da \xrightarrow[\text{theorem}]{\text{Stokes'}} \vec{\nabla} \times \vec{E} = -\frac{\partial \vec{B}}{\partial t}.$$

Ampere–Maxwell law:

$$\oint_C \vec{B} \circ d\vec{l} = \mu_0 \left(I_{enc} + \varepsilon_0 \frac{d}{dt} \int_S \vec{E} \circ \hat{n}\, da \right) \xrightarrow[\text{theorem}]{\text{Stokes'}} \vec{\nabla} \times \vec{B} = \mu_0 \left(\vec{J} + \varepsilon_0 \frac{\partial \vec{E}}{\partial t} \right).$$

Along with the divergence theorem and Stokes' theorem, you'll also find a discussion of the gradient operator and some useful vector identities in this chapter. In addition, since the goal is to arrive at the wave equation, here are the expanded views of the wave equation for electric and magnetic fields:

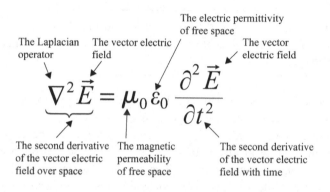

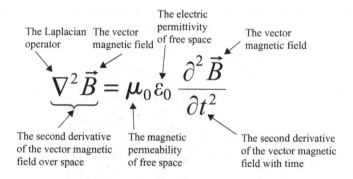

$$\left| \oint_S \vec{A} \circ \hat{n} \, da = \int_V (\vec{\nabla} \circ \vec{A}) \, dV \right|$$ **The divergence theorem**

The divergence theorem is a vector–calculus relation that equates the flux of a vector field to the volume integral of the field's divergence. The relationships between line, surface, and volume integrals were explored by several of the leading mathematical thinkers of the eighteenth and nineteenth centuries, including J. L. LaGrange in Italy, M. V. Ostrogradsky in Russia, G. Green in England, and C. F. Gauss in Germany. In some texts, you'll find the divergence theorem referred to as "Gauss's theorem" (which you should not confuse with Gauss's law).

The divergence theorem may be stated as follows:

> The flux of a vector field through a closed surface S is equal to the integral of the divergence of that field over a volume V for which S is a boundary.

This theorem applies to vector fields that are "smooth" in the sense of being continuous and having continuous derivatives.

To understand the physical basis for the divergence theorem, recall that the divergence at any point is defined as the flux through a small surface surrounding that point divided by the volume enclosed by that surface as it shrinks to zero. Now consider the flux through the cubical cells within the volume V shown in Figure 5.1.

For interior cells (those not touching the surface of V), the faces are shared with six adjacent cells (only some of which are shown in Figure 5.1 for clarity). For each shared face, the positive (outward) flux from one

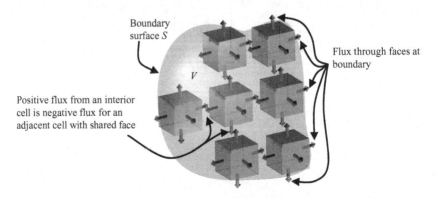

Boundary surface S

Flux through faces at boundary

V

Positive flux from an interior cell is negative flux for an adjacent cell with shared face

Figure 5.1 Cubical cells within volume V bounded by surface S.

cell is identical in amplitude and opposite in sign to the negative (inward) flux of the adjacent cell over that same face. Since all interior cells share faces with adjacent cells, only those faces that lie along the boundary surface S of volume V contribute to the flux through S.

This means that adding the flux through all the faces of all the cells throughout volume V leaves only the flux through the bounding surface S. Moreover, in the limit of infinitesimally small cells, the definition of divergence tells you that the divergence of the vector field at any point is the outward flux from that point. So, adding the flux of each cell is the same as integrating the divergence over the entire volume. Thus,

$$\oint_S \vec{A} \circ \hat{n} \, da = \int_V (\vec{\nabla} \circ \vec{A}) \, dV. \tag{5.1}$$

This is the divergence theorem – the integral of the divergence of a vector field over V is identical to the flux through S. And how is this useful? For one thing, it can get you from the integral form to the differential form of Gauss's laws. In the case of electric fields, the integral form of Gauss's law is

$$\oint_S \vec{E} \circ \hat{n} \, da = q_{\text{enc}}/\varepsilon_0.$$

Or, since the enclosed charge is the volume integral of the charge density ρ,

$$\oint_S \vec{E} \circ \hat{n} \, da = \frac{1}{\varepsilon_0} \int_V \rho \, dV.$$

Now, apply the divergence theorem to the left side of Gauss's law,

$$\oint_S \vec{E} \circ \hat{n} \, da = \int_V \vec{\nabla} \circ \vec{E} \, dV = \frac{1}{\varepsilon_0} \int_V \rho \, dV = \int_V \frac{\rho}{\varepsilon_0} \, dV.$$

Since this equality must hold for all volumes, the integrands must be equal. Thus,

$$\vec{\nabla} \circ \vec{E} = \rho/\varepsilon_0,$$

which is the differential form of Gauss's law for electric fields. The same approach applied to the integral form of Gauss's law for magnetic fields leads to

$$\vec{\nabla} \circ \vec{B} = 0$$

as you might expect.

$$\oint_C \vec{A} \circ d\vec{l} = \int_S (\vec{\nabla} \times \vec{A}) \circ \hat{n} \, da$$ **Stokes' theorem**

Whereas the divergence theorem relates a surface integral to a volume integral, Stokes' theorem relates a line integral to a surface integral. William Thompson (later Lord Kelvin) referred to this relation in a letter in 1850, and it was G. G. Stokes who made it famous by setting its proof as an exam question for students at Cambridge. You may encounter generalized statements of Stokes' theorem, but the form relevant to Maxwell's Equations (sometimes called the "Kelvin–Stokes theorem") may be stated as follows:

> The circulation of a vector field over a closed path C is equal to the integral of the normal component of the curl of that field over a surface S for which C is a boundary.

This theorem applies to vector fields that are "smooth" in the sense of being continuous and having continuous derivatives.

The physical basis for Stokes' theorem may be understood by recalling that the curl at any point is defined as the circulation around a small path surrounding that point divided by the surface area enclosed by that path as it shrinks to zero. Consider the circulation around the small squares on the surface S shown in Figure 5.2.

For interior squares (those not touching the edge of the surface), each edge is shared with an adjacent square. For each shared edge, the circulation from one square is identical in amplitude and opposite in sign to the

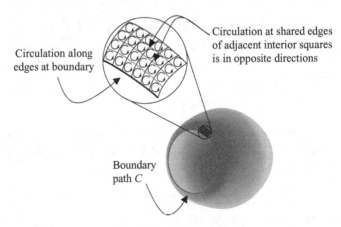

Circulation along edges at boundary

Circulation at shared edges of adjacent interior squares is in opposite directions

Boundary path C

Figure 5.2 Squares on surface S bounded by path C.

circulation of the adjacent square over that same edge. It is only those edges that lie along the boundary path C of surface S that are not shared with an adjacent square, and which contribute to the circulation around C.

Thus, adding the circulation around all the edges of all the squares over surface S leaves only the circulation around the bounding path C. In addition, in the limit of infinitesimally small squares, the definition of curl tells you that adding the circulation of each square is the same as integrating the normal component of the curl of the vector field over the surface. So,

$$\oint_C \vec{A} \circ d\vec{l} = \int_S (\vec{\nabla} \times \vec{A}) \circ \hat{n} \, da. \qquad (5.2)$$

Stokes' theorem does for line integrals and the curl what the divergence theorem does for surface integrals and the divergence. In this case, the integral of the normal component of the curl over S is identical to the circulation around C. Moreover, just as the divergence theorem leads from the integral to the differential form of Gauss's laws, Stokes' theorem can be applied to the integral form of Faraday's law and the Ampere–Maxwell law.

Consider the integral form of Faraday's law, which relates the circulation of the electric field around a path C to the change in magnetic flux through a surface S for which C is a boundary,

$$\oint_C \vec{E} \circ d\vec{l} = -\frac{d}{dt} \int_S \vec{B} \circ \hat{n} \, da.$$

Applying Stokes' theorem to the circulation on the left side gives

$$\oint_C \vec{E} \circ d\vec{l} = \int_S (\vec{\nabla} \times \vec{E}) \circ \hat{n} \, da$$

Thus, Faraday's law becomes

$$\int_S (\vec{\nabla} \times \vec{E}) \circ \hat{n} \, da = -\frac{d}{dt} \int_S \vec{B} \circ \hat{n} \, da.$$

For stationary geometries, the time derivative can be moved inside the integral, so this is

$$\int_S (\vec{\nabla} \times \vec{E}) \circ \hat{n} \, da = \int_S \left(-\frac{\partial \vec{B}}{\partial t} \circ \hat{n} \right) da,$$

where the partial derivative indicates that the magnetic field may change over space as well as time. Since this equality must hold over all surfaces, the integrands must be equal, giving

$$\vec{\nabla} \times \vec{E} = -\frac{\partial \vec{B}}{\partial t},$$

which is the differential form of Faraday's law, relating the curl of the electric field at a point to the time rate of change of the magnetic field at that point.

Stokes' theorem may also be used to find the differential form of the Ampere–Maxwell law. Recall that the integral form relates the circulation of the magnetic field around a path C to the current enclosed by that path and the time rate of change of electric flux through a surface S bound by path C:

$$\oint_C \vec{B} \circ d\vec{l} = \mu_0 \left(I_{\text{enc}} + \varepsilon_0 \frac{d}{dt} \int_S \vec{E} \circ \hat{n}\, da \right).$$

Applying Stokes' theorem to the circulation gives

$$\oint_C \vec{B} \circ d\vec{l} = \int_S (\vec{\nabla} \times \vec{B}) \circ \hat{n}\, da,$$

which makes the Ampere–Maxwell law

$$\int_S (\vec{\nabla} \times \vec{B}) \circ \hat{n}\, da = \mu_0 \left(I_{\text{enc}} + \varepsilon_0 \frac{d}{dt} \int_S \vec{E} \circ \hat{n}\, da \right).$$

The enclosed current may be written as the integral of the normal component of the current density

$$I_{\text{enc}} = \int_S \vec{J} \circ \hat{n}\, da,$$

and the Ampere–Maxwell law becomes

$$\int_S (\vec{\nabla} \times \vec{B}) \circ \hat{n}\, da = \mu_0 \left(\int_S \vec{J} \circ \hat{n}\, da + \int_S \varepsilon_0 \frac{\partial \vec{E}}{\partial t} \circ \hat{n}\, da \right).$$

Once again, for this equality to hold over all surfaces, the integrands must be equal, meaning

$$\vec{\nabla} \times \vec{B} = \mu_0 \left(\vec{J} + \varepsilon_0 \frac{\partial \vec{E}}{\partial t} \right).$$

This is the differential form of the Ampere–Maxwell law, relating the curl of the magnetic field at a point to the current density and time rate of change of the electric field at that point.

$\boxed{\vec{\nabla}()}$ The gradient

To understand how Maxwell's Equations lead to the wave equation, it is necessary to comprehend a third differential operation used in vector calculus – the gradient. Similar to the divergence and the curl, the gradient involves partial derivatives taken in three orthogonal directions. However, whereas the divergence measures the tendency of a vector field to flow away from a point and the curl indicates the circulation of a vector field around a point, the gradient applies to *scalar fields*. Unlike a vector field, a scalar field is specified entirely by its magnitude at various locations: one example of a scalar field is the height of terrain above sea level.

What does the gradient tell you about a scalar field? Two important things: the magnitude of the gradient indicates how quickly the field is changing over space, and the direction of the gradient indicates the direction in that the field is changing most quickly with distance.

Therefore, although the gradient operates on a scalar field, the result of the gradient operation is a vector, with both magnitude and direction. Thus, if the scalar field represents terrain height, the magnitude of the gradient at any location tells you how steeply the ground is sloped at that location, and the direction of the gradient points *uphill* along the steepest slope.

The definition of the gradient of the scalar field ψ is

$$\text{grad}(\psi) = \vec{\nabla}\psi \equiv \hat{i}\frac{\partial \psi}{\partial x} + \hat{j}\frac{\partial \psi}{\partial y} + \hat{k}\frac{\partial \psi}{\partial z} \qquad \text{(Cartesian).} \qquad (5.3)$$

Thus, the x-component of the gradient of ψ indicates the slope of the scalar field in the x-direction, the y-component indicates the slope in the y-direction, and the z-component indicates the slope in the z-direction. The square root of the sum of the squares of these components provides the total steepness of the slope at the location at which the gradient is taken.

In cylindrical and spherical coordinates, the gradient is

$$\vec{\nabla}\psi \equiv \hat{r}\frac{\partial \psi}{\partial r} + \hat{\varphi}\frac{1}{r}\frac{\partial \psi}{\partial \varphi} + \hat{z}\frac{\partial \psi}{\partial z} \qquad \text{(cylindrical)} \qquad (5.4)$$

and

$$\vec{\nabla}\psi \equiv \hat{r}\frac{\partial \psi}{\partial r} + \hat{\theta}\frac{1}{r}\frac{\partial \psi}{\partial \theta} + \hat{\varphi}\frac{1}{r\sin\theta}\frac{\partial \psi}{\partial \varphi} \qquad \text{(spherical).} \qquad (5.5)$$

$\boxed{\vec{\nabla}, \vec{\nabla}\circ, \vec{\nabla}\times}$ **Some useful identities**

Here is a quick review of the del differential operator and its three uses relevant to Maxwell's Equations:

Del:

$$\vec{\nabla} \equiv \hat{i}\frac{\partial}{\partial x} + \hat{j}\frac{\partial}{\partial y} + \hat{k}\frac{\partial}{\partial z}$$

> Del (nabla) represents a multipurpose differential operator that can operate on scalar or vector fields and produce scalar or vector results.

Gradient:

$$\vec{\nabla}\psi \equiv \hat{i}\frac{\partial \psi}{\partial x} + \hat{j}\frac{\partial \psi}{\partial y} + \hat{k}\frac{\partial \psi}{\partial z}$$

> The gradient operates on a scalar field and produces a vector result that indicates the rate of spatial change of the field at a point and the direction of steepest increase from that point.

Divergence:

$$\vec{\nabla}\circ\vec{A} \equiv \frac{\partial A_x}{\partial x} + \frac{\partial A_y}{\partial y} + \frac{\partial A_z}{\partial z}$$

> The divergence operates on a vector field and produces a scalar result that indicates the tendency of the field to flow away from a point.

Curl:

$$\vec{\nabla}\times\vec{A} \equiv \left(\frac{\partial A_z}{\partial y} - \frac{\partial A_y}{\partial z}\right)\hat{i} + \left(\frac{\partial A_x}{\partial z} - \frac{\partial A_z}{\partial x}\right)\hat{j} + \left(\frac{\partial A_y}{\partial x} - \frac{\partial A_x}{\partial y}\right)\hat{k}$$

> The curl operates on a vector field and produces a vector result that indicates the tendency of the field to circulate around a point and the direction of the axis of greatest circulation.

Once you're comfortable with the meaning of each of these operators, you should be aware of several useful relations between them (note that the following relations apply to fields that are continuous and that have continuous derivatives).

The curl of the gradient of any scalar field is zero.

$$\vec{\nabla} \times \vec{\nabla}\psi = 0, \tag{5.6}$$

which you may readily verify by taking the appropriate derivatives.

Another useful relation involves the divergence of the gradient of a scalar field; this is called the Laplacian of the field:

$$\vec{\nabla} \circ \vec{\nabla}\psi = \nabla^2\psi = \frac{\partial^2\psi}{\partial x^2} + \frac{\partial^2\psi}{\partial y^2} + \frac{\partial^2\psi}{\partial z^2} \qquad \text{(Cartesian)}. \tag{5.7}$$

The usefulness of these relations can be illustrated by applying them to the electric field as described by Maxwell's Equations. Consider, for example, the fact that the curl of the electrostatic field is zero (since electric field lines diverge from positive charge and converge upon negative charge, but do not circulate back upon themselves). Equation 5.6 indicates that as a curl-free (irrotational) field, the electrostatic field $\vec{E}$ may be treated as the gradient of another quantity called the scalar potential V:

$$\vec{E} = -\vec{\nabla}V, \tag{5.8}$$

where the minus sign is needed because the gradient points toward the greatest *increase* in the scalar field, and by convention the electric force on a positive charge is toward *lower* potential. Now apply the differential form of Gauss's law for electric fields:

$$\vec{\nabla} \circ \vec{E} = \frac{\rho}{\varepsilon_0},$$

which, combined with Equation 5.8, gives

$$\nabla^2 V = -\frac{\rho}{\varepsilon_0}. \tag{5.9}$$

This is called Poisson's equation, and it is often the best way to find the electrostatic field when you are not able to construct a special Gaussian surface. In such cases, it may be possible to solve Poisson's Equation for the electric potential V and then determine $\vec{E}$ by taking the gradient of the potential.

$$\boxed{\nabla^2 \vec{A} = \frac{1}{v^2}\frac{\partial^2 \vec{A}}{\partial t^2}}\ \text{The wave equation}$$

With the differential form of Maxwell's Equations and several vector operator identities in hand, the trip to the wave equation is a short one. Begin by taking the curl of both sides of the differential form of Faraday's law

$$\vec{\nabla} \times (\vec{\nabla} \times \vec{E}) = \vec{\nabla} \times \left(-\frac{\partial \vec{B}}{\partial t}\right) = -\frac{\partial(\vec{\nabla} \times \vec{B})}{\partial t}. \qquad (5.10)$$

Notice that the curl and time derivatives have been interchanged in the final term; as in previous sections, the fields are assumed to be sufficiently smooth to permit this.

Another useful vector operator identity says that the curl of the curl of any vector field equals the gradient of the divergence of the field minus the Laplacian of the field:

$$\vec{\nabla} \times (\vec{\nabla} \times \vec{A}) = \vec{\nabla}(\vec{\nabla} \circ \vec{A}) - \nabla^2 \vec{A}. \qquad (5.11)$$

This relation uses a vector version of the Laplacian operator that is constructed by applying the Laplacian to the components of a vector field:

$$\nabla^2 \vec{A} = \nabla^2 A_x \hat{i} + \nabla^2 A_y \hat{j} + \nabla^2 A_z \hat{k} \qquad \text{(Cartesian)}. \qquad (5.12)$$

Thus,

$$\vec{\nabla} \times (\vec{\nabla} \times \vec{E}) = \vec{\nabla}(\vec{\nabla} \circ \vec{E}) - \nabla^2 \vec{E} = -\frac{\partial(\vec{\nabla} \times \vec{B})}{\partial t}. \qquad (5.13)$$

However, you know the curl of the magnetic field from the differential form of the Ampere–Maxwell law:

$$\vec{\nabla} \times \vec{B} = \mu_0 \left(\vec{J} + \varepsilon_0 \frac{\partial \vec{E}}{\partial t}\right).$$

So,

$$\vec{\nabla} \times (\vec{\nabla} \times \vec{E}) = \vec{\nabla}(\vec{\nabla} \circ \vec{E}) - \nabla^2 \vec{E} = -\frac{\partial\left[\mu_0\left(\vec{J} + \varepsilon_0(\partial \vec{E}/\partial t)\right)\right]}{\partial t}.$$

This looks difficult, but one simplification can be achieved using Gauss's law for electric fields:

$$\vec{\nabla} \circ \vec{E} = \frac{\rho}{\varepsilon_0},$$

which means

$$\vec{\nabla} \times (\vec{\nabla} \times \vec{E}) = \vec{\nabla}\left(\frac{\rho}{\varepsilon_0}\right) - \nabla^2 \vec{E} = -\frac{\partial[\mu_0(\vec{J} + \varepsilon_0(\partial\vec{E}/\partial t))]}{\partial t}$$

$$= -\mu_0\frac{\partial\vec{J}}{\partial t} - \mu_0\varepsilon_0\frac{\partial^2\vec{E}}{\partial t^2}.$$

Gathering terms containing the electric field on the left side of this equation gives

$$\nabla^2\vec{E} - \mu_0\varepsilon_0\frac{\partial^2\vec{E}}{\partial t^2} = \vec{\nabla}\left(\frac{\rho}{\varepsilon_0}\right) + \mu_0\frac{\partial\vec{J}}{\partial t}.$$

In a charge- and current-free region, $\rho = 0$ and $\vec{J} = 0$, so

$$\nabla^2\vec{E} = \mu_0\varepsilon_0\frac{\partial^2\vec{E}}{\partial t^2}. \tag{5.14}$$

This is a linear, second-order, homogeneous partial differential equation that describes an electric field that travels from one location to another – in short, a propagating wave. Here is a quick reminder of the meaning of each of the characteristics of the wave equation:

Linear: The time and space derivatives of the wave function ($\vec{E}$ in this case) appear to the first power and without cross terms.

Second-order: The highest derivative present is the second derivative.

Homogeneous: All terms involve the wave function or its derivatives, so no forcing or source terms are present.

Partial: The wave function is a function of multiple variables (space and time in this case).

A similar analysis beginning with the curl of both sides of the Ampere–Maxwell law leads to

$$\nabla^2\vec{B} = \mu_0\varepsilon_0\frac{\partial^2\vec{B}}{\partial t^2}, \tag{5.15}$$

which is identical in form to the wave equation for the electric field.

This form of the wave equation doesn't just tell you that you have a wave – it provides the velocity of propagation as well. It is right there in the constants multiplying the time derivative, because the general form of the wave equation is this

$$\nabla^2 \vec{A} = \frac{1}{v^2} \frac{\partial^2 \vec{A}}{\partial t^2},$$ (5.16)

where v is the speed of propagation of the wave. Thus, for the electric and magnetic fields

$$\frac{1}{v^2} = \mu_0 \varepsilon_0,$$

or

$$v = \sqrt{\frac{1}{\mu_0 \varepsilon_0}}.$$ (5.17)

Inserting values for the magnetic permeability and electric permittivity of free space,

$$v = \sqrt{\frac{1}{[4\pi \times 10^{-7} \mathrm{m\ kg/C^2}][8.8541878 \times 10^{-12}\ \mathrm{C^2\ s^2/kg\ m^3}]}},$$

or

$$v = \sqrt{8.987552{\times}10^{16}\ \mathrm{m^2/s^2}} = 2.9979 \times 10^8\ \mathrm{m/s}.$$

It was the agreement of the calculated velocity of propagation with the measured speed of light that caused Maxwell to write, "light is an electromagnetic disturbance propagated through the field according to electromagnetic laws."

Appendix: Maxwell's Equations in matter

Maxwell's Equations as presented in Chapters 1–4 apply to electric and magnetic fields in matter as well as in free space. However, when you're dealing with fields inside matter, remember the following points:

- The enclosed charge in the integral form of Gauss's law for electric fields (and current density in the differential form) includes ALL charge – bound as well as free.
- The enclosed current in the integral form of the Ampere–Maxwell law (and volume current density in the differential form) includes ALL currents – bound and polarization as well as free.

Since the bound charge may be difficult to determine, in this Appendix you'll find versions of the differential and integral forms of Gauss's law for electric fields that depend only on the free charge. Likewise, you'll find versions of the differential and integral form of the Ampere–Maxwell law that depend only on the free current.

What about Gauss's law for magnetic fields and Faraday's law? Since those laws don't directly involve electric charge or current, there's no need to derive more "matter friendly" versions of them.

Gauss's law for electric fields: Within a dielectric material, positive and negative charges may become slightly displaced when an electric field is applied. When a positive charge Q is separated by distance s from an equal negative charge $-Q$, the electric "dipole moment" is given by

$$\vec{p} = Q\vec{s}, \tag{A.1}$$

where $\vec{s}$ is a vector directed from the negative to the positive charge with magnitude equal to the distance between the charges. For a dielectric

125

material with N molecules per unit volume, the dipole moment per unit volume is

$$\vec{P} = N\vec{p},\qquad\text{(A.2)}$$

a quantity which is also called the "electric polarization" of the material. If the polarization is uniform, bound charge appears only on the surface of the material. But if the polarization varies from point to point within the dielectric, there are accumulations of charge within the material, with volume charge density given by

$$\rho_b = -\vec{\nabla} \circ \vec{P},\qquad\text{(A.3)}$$

where ρ_b represents the volume density of bound charge (charge that's displaced by the electric field but does not move freely through the material).

What is the relevance of this to Gauss's law for electric fields? Recall that in the differential form of Gauss's law, the divergence of the electric field is

$$\vec{\nabla} \circ \vec{E} = \frac{\rho}{\varepsilon_0},$$

where ρ is the total charge density. Within matter, the total charge density consists of both free and bound charge densities:

$$\rho = \rho_f + \rho_b,\qquad\text{(A.4)}$$

where ρ is the total charge density, ρ_f is the free charge density, and ρ_b is the bound charge density. Thus, Gauss's law may be written as

$$\vec{\nabla} \circ \vec{E} = \frac{\rho}{\varepsilon_0} = \frac{\rho_f + \rho_b}{\varepsilon_0}.\qquad\text{(A.5)}$$

Substituting the negative divergence of the polarization for the bound charge and multiplying through by the permittivity of free space gives

$$\vec{\nabla} \circ \varepsilon_0 \vec{E} = \rho_f + \rho_b = \rho_f - \vec{\nabla} \circ \vec{P},\qquad\text{(A.6)}$$

or

$$\vec{\nabla} \circ \varepsilon_0 \vec{E} + \vec{\nabla} \circ \vec{P} = \rho_f.\qquad\text{(A.7)}$$

Collecting terms within the divergence operator gives

$$\vec{\nabla} \circ (\varepsilon_0 \vec{E} + \vec{P}) = \rho_f.\qquad\text{(A.8)}$$

In this form of Gauss's law, the term in parentheses is often written as a vector called the "displacement," which is defined as

$$\vec{D} = \varepsilon_0 \vec{E} + \vec{P}. \tag{A.9}$$

Substituting this expression into equation (A.8) gives

$$\vec{\nabla} \circ \vec{D} = \rho_f, \tag{A.10}$$

which is a version of the differential form of Gauss's law that depends only on the density of free charge.

Using the divergence theorem gives the integral form of Gauss's law for electric fields in terms of the flux of the displacement and enclosed free charge:

$$\oint_S \vec{D} \circ \hat{n} \, da = q_{\text{free, enc}}. \tag{A.11}$$

What is the physical significance of the displacement $\vec{D}$? In free space, the displacement is a vector field proportional to the electric field – pointing in the same direction as $\vec{E}$ and with magnitude scaled by the vacuum permittivity. But in polarizable matter, the displacement field may differ significantly from the electric field. You should note, for example, that the displacement is not necessarily irrotational – it will have curl if the polarization does, as can be seen by taking the curl of both sides of Equation A.9.

The usefulness of $\vec{D}$ comes about in situations for which the free charge is known and for which symmetry considerations allow you to extract the displacement from the integral of Equation A.11. In those cases, you may be able to determine the electric field within a linear dielectric material by finding $\vec{D}$ on the basis of the free charge and then dividing by the permittivity of the medium to find the electric field.

The Ampere–Maxwell law: Just as applied electric fields induce polarization (electric dipole moment per unit volume) within dielectrics, applied magnetic fields induce "magnetization" (magnetic dipole moment per unit volume) within magnetic materials. And just as bound electric charges act as the source of additional electric fields within the material, bound currents may act as the source of additional magnetic fields. In that case, the bound current density is given by the curl of the magnetization:

$$\vec{J}_b = \vec{\nabla} \times \vec{M}. \tag{A.12}$$

where $\vec{J}_b$ is the bound current density and $\vec{M}$ represents the magnetization of the material.

Another contribution to the current density within matter comes from the time rate of change of the polarization, since any movement of charge constitutes an electric current. The polarization current density is given by

$$\vec{J}_P = \frac{\partial \vec{P}}{\partial t}.$$
(A.13)

Thus, the total current density includes not only the free current density, but the bound and polarization current densities as well:

$$\vec{J} = \vec{J}_f + \vec{J}_b + \vec{J}_P.$$
(A.14)

Thus, the Ampere–Maxwell law in differential form may be written as

$$\vec{\nabla} \times \vec{B} = \mu_0 \left(\vec{J}_f + \vec{J}_b + \vec{J}_P + \varepsilon_0 \frac{\partial \vec{E}}{\partial t} \right).$$
(A.15)

Inserting the expressions for the bound and polarization current and dividing by the permeability of free space

$$\frac{1}{\mu_0} \vec{\nabla} \times \vec{B} = \vec{J}_f + \vec{\nabla} \times \vec{M} + \frac{\partial \vec{P}}{\partial t} + \varepsilon_0 \frac{\partial \vec{E}}{\partial t}.$$
(A.16)

Gathering curl terms and time-derivative terms gives

$$\vec{\nabla} \times \frac{\vec{B}}{\mu_0} - \vec{\nabla} \times \vec{M} = \vec{J}_f + \frac{\partial \vec{P}}{\partial t} + \frac{\partial(\varepsilon_0 \vec{E})}{\partial t}.$$
(A.17)

Moving the terms inside the curl and derivative operators makes this

$$\vec{\nabla} \times \left(\frac{\vec{B}}{\mu_0} - \vec{M} \right) = \vec{J}_f + \frac{\partial(\varepsilon_0 \vec{E} + \vec{P})}{\partial t}.$$
(A.18)

In this form of the Ampere–Maxwell law, the term in parentheses on the left side is written as a vector sometimes called the "magnetic field intensity" or "magnetic field strength" and defined as

$$\vec{H} = \frac{\vec{B}}{\mu_0} - \vec{M}.$$
(A.19)

Thus, the differential form of the Ampere–Maxwell law in terms of $\vec{H}$, $\vec{D}$, and the free current density is

$$\vec{\nabla} \times \vec{H} = \vec{J}_{\text{free}} + \frac{\partial \vec{D}}{\partial t}. \tag{A.20}$$

Using Stokes' theorem gives the integral form of the Ampere–Maxwell law:

$$\oint_C \vec{H} \circ d\vec{l} = I_{\text{free, enc}} + \frac{d}{dt} \int_S \vec{D} \circ \hat{n} \, da \tag{A.21}$$

What is the physical significance of the magnetic intensity $\vec{H}$? In free space, the intensity is a vector field proportional to the magnetic field – pointing in the same direction as $\vec{B}$ and with magnitude scaled by the vacuum permeability. But just as $\vec{D}$ may differ from $\vec{E}$ inside dielectric materials, $\vec{H}$ may differ significantly from $\vec{B}$ in magnetic matter. For example, the magnetic intensity is not necessarily solenoidal – it will have divergence if the magnetization does, as can be seen by taking the divergence of both sides of Equation A.19.

As is the case for electric displacement, the usefulness of $\vec{H}$ comes about in situations for which you know the free current and for which symmetry considerations allow you to extract the magnetic intensity from the integral of Equation A.21. In such cases, you may be able to determine the magnetic field within a linear magnetic material by finding $\vec{H}$ on the basis of free current and then multiplying by the permeability of the medium to find the magnetic field.

Here is a summary of the integral and differential forms of all of Maxwell's Equations in matter:

Gauss's law for electric fields:

$$\oint_S \vec{D} \circ \hat{n}\, da = q_{\text{free, enc}} \qquad \text{(integral form)},$$

$$\vec{\nabla} \circ \vec{D} = \rho_{\text{free}} \qquad \text{(differential form)}.$$

Gauss's law for magnetic fields:

$$\oint_S \vec{B} \circ \hat{n}\, da = 0 \qquad \text{(integral form)},$$

$$\vec{\nabla} \circ \vec{B} = 0 \qquad \text{(differential form)}.$$

Faraday's law:

$$\oint_C \vec{E} \circ d\vec{l} = -\frac{d}{dt}\int_S \vec{B} \circ \hat{n}\, da \qquad \text{(integral form)},$$

$$\vec{\nabla} \times \vec{E} = -\frac{\partial \vec{B}}{\partial t} \qquad \text{(differential form)}.$$

Ampere–Maxwell law:

$$\oint_C \vec{H} \circ d\vec{l} = I_{\text{free, enc}} + \frac{d}{dt}\int_S \vec{D} \circ \hat{n}\, da \qquad \text{(integral form)},$$

$$\vec{\nabla} \times \vec{H} = \vec{J}_{\text{free}} + \frac{\partial \vec{D}}{\partial t} \qquad \text{(differential form)}.$$

Further reading

If you're looking for a comprehensive treatment of electricity and magnetism, you have several excellent texts from which to choose. Here are some that you may find useful:

Cottingham W. N. and Greenwood D. A., *Electricity and Magnetism*. Cambridge University Press, 1991; A concise survey of a wide range of topics in electricity and magnetism.

Griffiths, D. J., *Introduction to Electrodynamics*. Prentice-Hall, New Jersey, 1989; The standard undergraduate text at the intermediate level, with clear explanations and informal style.

Jackson, J. D., *Classical Electrodynamics*. Wiley & Sons, New York, 1998; The standard graduate text, but you must be solidly prepared before embarking.

Lorrain, P., Corson, D., and Lorrain, F., *Electromagnetic Fields and Waves*. Freeman, New York, 1988; Another excellent intermediate-level text, with detailed explanations supported by helpful diagrams.

Purcell, E. M., *Electricity and Magnetism Berkeley Physics Course, Vol. 2*. McGraw-Hill, New York, 1965; Probably the best of the introductory-level texts; elegantly written and carefully illustrated.

Wangsness, R. K., *Electromagnetic Fields*. Wiley, New York, 1986; Also a great intermediate-level text, especially useful as preparation for Jackson.

And for a comprehensible introduction to vector operators, with many examples drawn from electrostatics, check out:
Schey, H. M., *Div, Grad, Curl, and All That*. Norton, New York, 1997.

Index

Printed in the United States
By Bookmasters